I0069969

The Tortoise and Achilles

using Geometry Expressions™

to investigate the infinite

Written by Larry Ottman

Haddon Heights High School, Haddon Heights, New Jersey

Illustrations by Regina Doris Ottman

Saltire Software, Inc.
Tigard, OR, USA
www.saltire.com
www.geometryexpressions.com

Copyright © Saltire Software 2010

ISBN: 1-882564-24-3

Limited Permission for Reproduction

© 2010 by Saltire Software, Inc. All rights reserved. Saltire grants the teacher who purchases *The Tortoise and Achilles : using Geometry Expressions to investigate the infinite,* the right to reproduce material for use in their individual classroom.

Saltire Software
P.O. Box 230755
Tigard, OR 97281-0755
http://www.geometryexpressions.com/
http://www.saltire.com/
support@saltire.com

Table of Contents

Foreword

I can remember having arguments like this as a young kid and either I was a bizarre child (which is highly likely), or this is a typical situation played out in many childhood arguments. The sense of triumph as one combatant first proposes the insurmountable obstacle of the infinite, introduces one of the most important ideas of mathematics, and perhaps unknowingly the larger questions of philosophy, science, literature, and even religion. The apparent victory above is first achieved by skipping the inevitability of this argument going on for a *very* long time. But this quickly gives way to the absurdity that there is indeed a number "larger" than infinity and the can of worms opened from that competitive impulse brings up issues that have been debated for thousands of years.

Calculus is the study of the infinite and since much of secondary mathematics is designed to prepare one for the study of calculus, wrestling with the ideas of the infinite, even if informally, is extremely important for a student's mathematical development. That is the purpose of this book.

Among the first to discuss these ideas was the Greek philosopher / mathematician Zeno, and not long after him, Archimedes came the closest to "discovering" Calculus without the tools of modern mathematics. Though none of

Zeno's actual writings survive, Aristotle recorded accounts of Zeno's thoughts on the infinite, time and space in what have come to be known as Zeno's Paradoxes. One of those specifically involves the idea of a race in which a slower runner is given a head start and investigates the possibilities of the faster runner "catching up". This has come to be known as "Achilles and the Tortoise". These characters are our hosts as we use modern software to investigate Archimedes methods, ideas of the infinite, and Zeno's Paradoxes in an introduction to Calculus, without using Calculus.

Lesson One

The following appears in Aristotle's *Physics* from 333 BC:

"The Achilles: That the swiftest racer can never overtake the slowest, if the slowest is given any start at all; because the slowest will have passed beyond his starting-point when the swiftest reaches it, and beyond the point he has then reached when the swiftest reaches it and so on..."[1]

[1]

This is the only written record we have of what has come to be known as Zeno's paradox of Achilles and the Tortoise. Mathematicians, philosophers, and quantum physicists are still arguing about it thousands of years later!

Achilles: Surely, you're not saying that I can't catch you in a race?

Tortoise: That's exactly what I am saying! And please, don't call me Shirley.

Achilles: Very funny, but I don't see how that is possible…

---

1 Mazur, Joseph. *Zeno's Paradox: unraveling the ancient mystery behind the science of space and time.* New York: Plume Publishing, 2008. p. 4.

Tortoise: Well, that's why they call it a paradox! We will get more into that later. First, we need to discuss some basic mathematics that is at the heart of this apparent contradiction.

A **sequence** is simply a listing of numbers. The most interesting sequences progress in some sort of predictable pattern. One such type of sequence is called a **geometric sequence**. If you divide any two successive terms of a geometric sequence you always get the same value, called a **common ratio**.

Geometric sequences can either be **finite**, meaning that we can count the number of terms, or **infinite**. In the case of the latter, there is no end to the sequence.

Before we can proceed, you need to be able to demonstrate some facility and understanding with geometric sequences.

Exciting Examples

1. Given the following sequence:

$$1, 2, 4, 8, \ldots$$

    a. Verify that this sequence is geometric and identify the common ratio.

    b. Is the sequence finite, or infinite?

2. Which of the following sequences are geometric? If it is geometric, identify the common ratio.

    a. $3, 6, 9, 12 \ldots$ b. $2, 4, 16, 128$

    c. $1, 3, 9, 27, 81 \ldots$ d. $\dfrac{1}{3}, \dfrac{1}{6}, \dfrac{1}{12}, \dfrac{1}{24} \ldots$

3. Identify which of the following sequences are geometric. If they are geometric, identify the common ratio.

a. $4, 6, 9, 13.5\ldots$

b. $18, 12, 8, \dfrac{16}{3}\ldots$

c. $1, -2, 4, -8\ldots$

d. $-54, 18, -6, 2\ldots$

4. Write the first 5 terms of the geometric sequence that starts with 12 and has a common ratio of -2/3.

5. Write the 8th term of the geometric sequence that starts with 5 and has a common ratio of -4.

Interesting Investigation

Tortoise: Ok! With some basics out of the way, let's use Geometry Expressions to understand a sequence in a more visual form.

Achilles: I could go for that. How about this sequence:

$$1 + \frac{1}{2} + \frac{1}{4} + \frac{1}{8} + \frac{1}{16} + \ldots$$

Tortoise: That's perfect. Let's represent this sequence geometrically as a progression of areas.

1. Open Geometry Expressions. Choose **FILE-NEW** to open a new Geometry Expressions (**GX** to its friends!) document. Save it as *lesson1.gx*.

2. Choose **DRAW-POLYGON** to construct a quadrilateral ABCD. Don't waste time trying to make it square as we will do that next!

> Like most good software, **GX** allows you to access commands using icons or menus. If the Draw Panel is visible, the icon for **DRAW-POLYGON** is [icon]. If the tool panel is not visible, go to **VIEW-TOOL PANEL CONFIGURATIONS** and choose **DEFAULT**. From now on in the text, we will use both the icons and the menu .commands.

3. Constrain the length of side **AB** to be one unit. With side **AB** selected, choose **CONSTRAIN-DISTANCE/LENGTH** ([icon]) and then type 1 and press enter.

4. Select sides **AB** and **BC** of your square (hold down the **SHIFT** key while selecting to select multiple objects), and choose **CONSTRAIN-CONGRUENT** ([icon]) to force them to be the same length. Notice that the selected sides have small congruence symbols.

5. Repeat this procedure by constraining the remaining two sides to be congruent to the original (or a side that has already been constrained).

6. Next, we need to also constrain the angles to be right. Select any two adjacent sides and choose **CONSTRAIN-PERPENDICULAR** ([icon]). Oh, did you get the conflicting constraints message?

7. It may seem natural to force *all* the angles to be perpendicular because, after all, a square is made up of four right angles! However, because we already constrained the sides to be congruent, when one pair of angles is right, all the vertices must be right. This is an important point for novice users of *Geometry Expressions* to understand. If you try to "over-constrain" a figure, *GX* will return an error message. You are either asking *GX* to do something that is already true, or that conflicts with something else you have constrained. However, in this case, there is only one pair of sides that *GX* will allow you to constrain. Select

Cancel in the message box and try to constrain another set of sides to be perpendicular.

8. When you have found the right set of sides to make perpendicular, check to see that the drawing is properly constrained. Try dragging and rotating the square by each of its endpoints and sides. You should observe that the square maintains its size and shape no matter how you drag it.

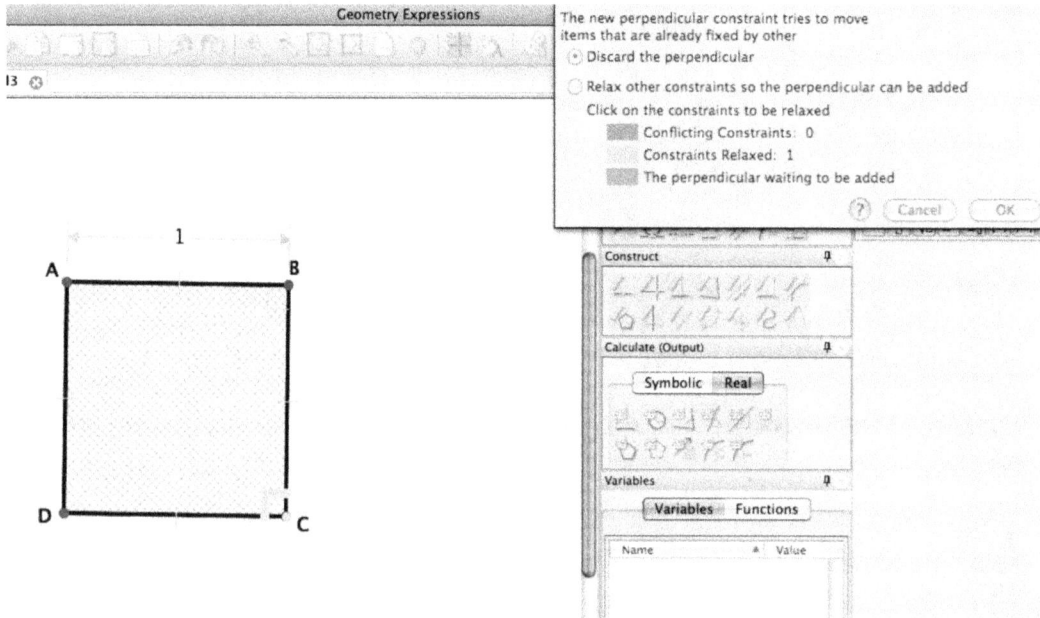

9. Let's now have *GX* calculate the area of the square. First, select the interior of the square.

GX can perform two types of measurements. It can give you a numeric answer, or it can calculate an algebraic expression for the measurement. Specify the type of measurement using the **CALCULATE** Tool Panel. In this case, choose **REAL.**

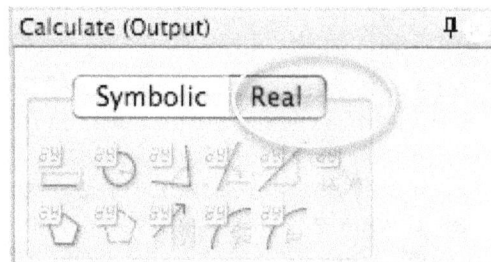

Notice if you choose **SYMBOLIC**, the icons change to indicate symbolic vs. real measurements.

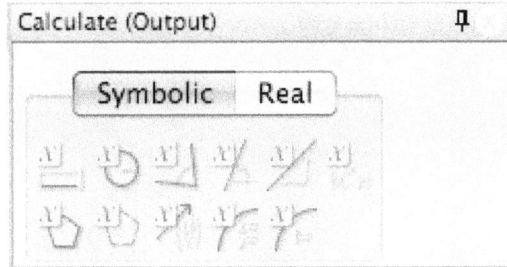

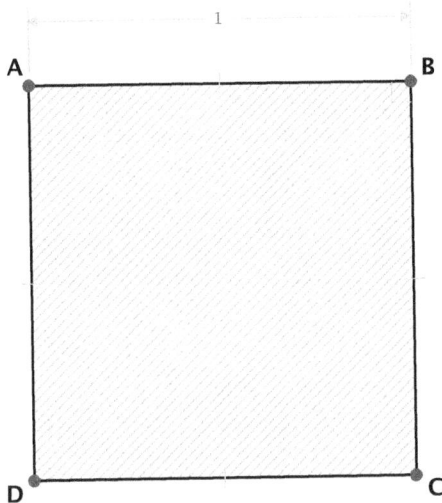

Calculate (Output)

Symbolic   Real

A

B

$z_0 \Rightarrow \sim 1$

Choose **REAL** and to measure the actual area. Real measurements will also have an approximation sign ($\sim$) next to them, indicating that they are measurements, and not mathematical calculations.

Output measurements and expressions in GX are, by default, notated using subscripts, in

D

C

this case, z_0.

We can easily change between the two types for an existing measurement by right-clicking (or control-click if you're missing a right mouse button) on the measurement and choosing **CONVERT TO (SYMBOLIC / REAL)**. Convert z_0 to **SYMBOLIC.**

10. Select the entire square (use **CNTRL-A** for select all) and make a duplicate using copy/paste (**CNTRL-C, CNTRL-V**).

11. Hide the interior of the square by selecting it and choosing **VIEW-HIDE (CNTRL-H)**.

12. Because the area of the original square is one, we can show the next term of

the sequence, $\dfrac{1}{2}$, as any number of bisections of the square. Let's choose to cut

it along the diagonal. Create a triangle using the **POLYGON** tool and connecting A, B, and D. Measure the symbolic area of triangle **ABD.**

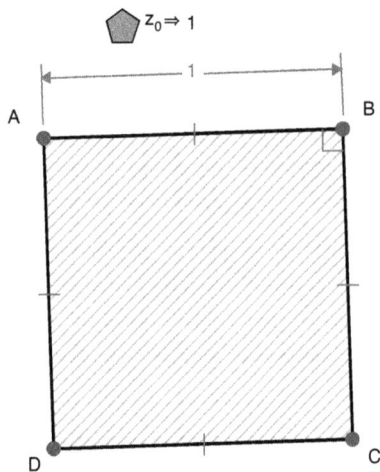

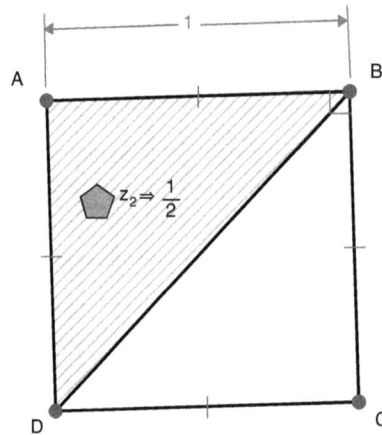

13. You may change the color and shading of the region using the **PREFERENCES**, or select the interior and use the right-click context menu if you wish.

14. Because the common ratio of the sequence is $\dfrac{1}{2}$, we can continue to repeat the pattern by slicing the empty part of the square in half and measuring the

resulting area. Select the diagonal **BD** and **CONSTRUCT** the **MIDPOINT (**) of the diagonal. Use the polygon tool to create and measure the area of the next triangle as follows:

$z_2 \Rightarrow \dfrac{1}{2}$

$z_3 \Rightarrow \dfrac{1}{4}$

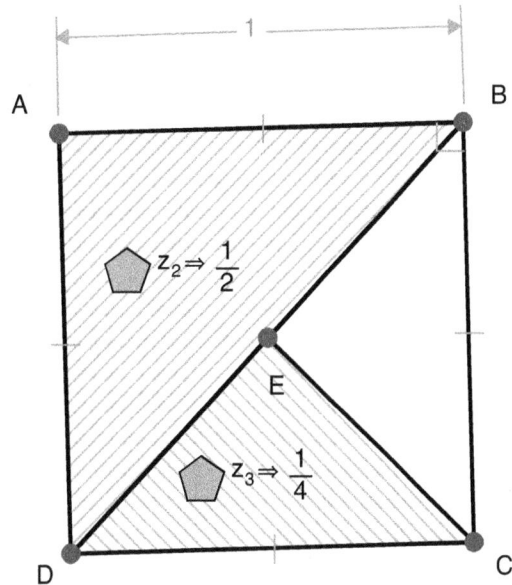

When you are using a **Drawing** tool in **GX**, that tool stays active until you choose a different tool. To exit the current tool, either choose the

select tool ( ), or press **ESC**. Also, when you do something wrong or mess up your drawing in a way that seems disastrous, remember: **CTRL-Z (UNDO)** is your best friend!

Tortoise: I thought *I* was your best friend?

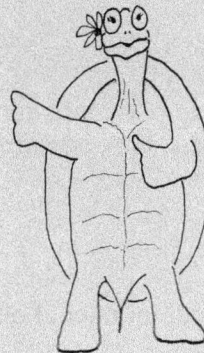

15. Complete two more iterations (iteration is a term we use in mathematics for a repetitive process) of the pattern. Make sure each time that you construct the midpoint of the diagonal of the empty triangle!

Tortoise: This is REALLY important!

To construct the midpoint of **BE,** hold the shift key (that's how we select more than one object) while selecting the endpoints **B** and **E**, and click **CONSTRUCT** the

MIDPOINT ($\frac{1}{2}$). The endpoint selection method will be necessary for any successive midpoints.

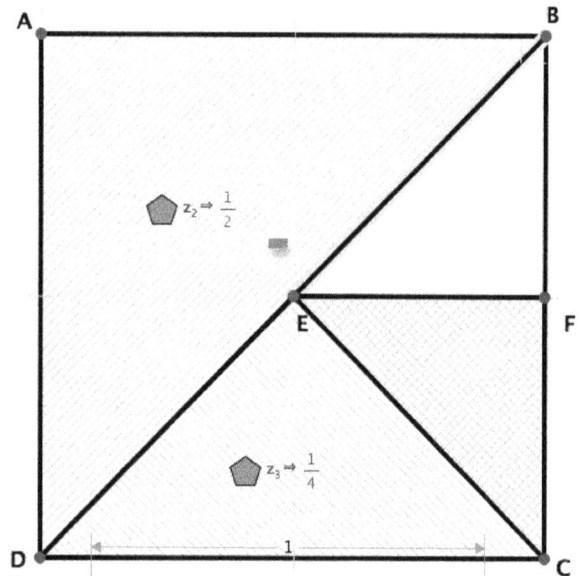

16. Complete one more iteration. Your final drawing should look something like this:

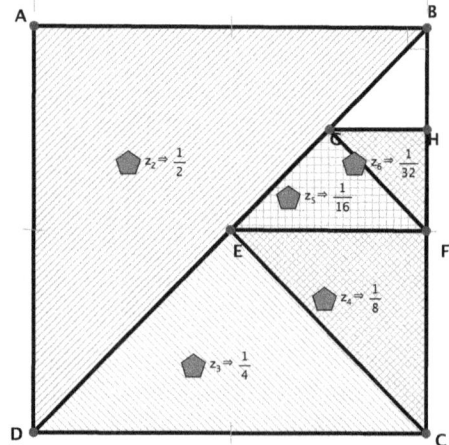

17. Have your teacher check your completed drawing.

TEACHER CHECK:_____

1. Describe the change in **each new** shaded region of the square as we continue the sequence.

2. What would be true of the area of **the new** shaded region if we were to repeat the process *many* times?

1. Now let's consider the same pattern as a **series**. In mathematics, a series is a sequence of sums.

$$1 + \frac{1}{2} + \frac{1}{4} + \frac{1}{8} + \frac{1}{16} + ...$$

2. We can represent the terms of the sequence with our successively smaller triangles and the total area as the sum of the series. First, let's convert all of our area measurements to **REAL**. You should remember how to do this!

3. Now, press $\boxed{x+y}$ and click in an empty portion of the window. This will be the expression z_7. Type z[0]+z[2]+z[3]+z[4]+z[5]+z[6] (the square bracket is GX's way of writing subscripts). If you are wondering why we skipped z[1], it is because that was the expression for the area of the total second square that we hid. The menu command for this is **DRAW-EXPRESSION**.

Write the value of your expression here:_____

4. Now, perform two more iterations of the series. Double-click on the expresion and add z_8 and z_9 to the sum. When you press *enter* the expressions will be recalculated.

Record the new value here:_____

More Quick Questions

1. Describe the change in the **total** shaded region as we continue the sequence.

2. What would be true of the area of **the total** shaded region if we were to repeat the process *many* times?

3. Complete the following table using your results.

Term/iteration number	New Shaded Area	Total Shaded Area
0		
1		
2		
3		
4		
5		
6		
7		

. . .

REALLY big number		
∞		

4. What does the zero term represent in the drawing?

5. **SAVE** your model and have your teacher check the final product.

   Teacher Check:_____

Lesson Two

Purpose: Derive the formula for the sum of an infinite geometric series using Zeno's Paradox.

Achilles: This is fun so far, but when do we get to the part where I beat you in a race?

Tortoise: Just hold on to your heels for a bit! We still have some work to do. Besides, I wouldn't be so sure of myself if I were you! Zeno wrote another paradox that I want to investigate first. Most people call it "The Dichotomy". It appears only in Aristotle's writing as:

   "That which is in locomotion must first arrive at the half-way stage before it arrives at the goal."

Achilles: So I'm over here right now, and if I want to walk over to you, I have to reach the halfway point first. I don't see a problem with that.

Tortoise: In order to reach this halfway point, you would first have to reach a point that is half of that distance, or a quarter. But to get there, you would have to reach a point half of that distance, and a point half of that distance, etc. Because you would continually have to cover half of these distances you would be doing this infinitely. Not only can you never walk over to me, but it is completely impossible for you to even get started!

Achilles: Well that's ridiculous! Look, I just walked over to you while you were talking - end of discussion!

Tortoise: This is what makes it a paradox! It is an apparent contradiction between what feels like reality and a method for explaining that reality. Fortunately for us, modern mathematics has a nice way of explaining this. Now get back over there. I'd like you to meet my friend the frog who will help me illustrate this paradox.

Frog: Hello, my baby!

Tortoise: Don't get me in trouble for copyright infringement please! Now, let's say that my frog is going to jump over to you. Imagine that froggy here is able to jump half the distance from me to you and then he continues to jump half of the remaining distances between us. The paradox is that he can never reach you because he will need to infinitely halve these distances. But it turns out that this is a geometric series and this infinite series of terms actually turns out to have a finite sum. *Geometry Expressions* can represent this nicely for us!

Achilles: Great, let's get started then!

Informative Instructions

1. Create a **NEW** GX drawing and save it as ***lesson2.gx***

2. Turn the coordinate axes on by pressing ![grid icon]. Use ![zoom icon] to zoom in on the interval between 0 and 1 and then ![hand icon] to adjust the positioning of the axes.

3. Think of our frog as starting one unit away from the origin and jumping half the distance to the origin. Half of some distance, no matter how small, is still some distance. However, as we get infinitely closer to the origin, this distance approaches 0. Place a starting point somewhere on the x-axis, but **not** on the origin.

4. Select the point and choose **CONSTRAIN-COORDINATES** (![icon]). Enter **(1,0)** in place of (x_0, y_0). This will represent the starting position of our frog.

5. Now, because he starts by jumping half the distance to the origin, choose ![icon] to draw a segment on the x-axis that extends to a new point **B** towards the origin from point **A**. Select this segment and choose **CONSTRAIN-DISTANCE (**

![icon] **).** Constrain the length of this segment to be $\dfrac{1}{2}$. Your drawing should look something like this:

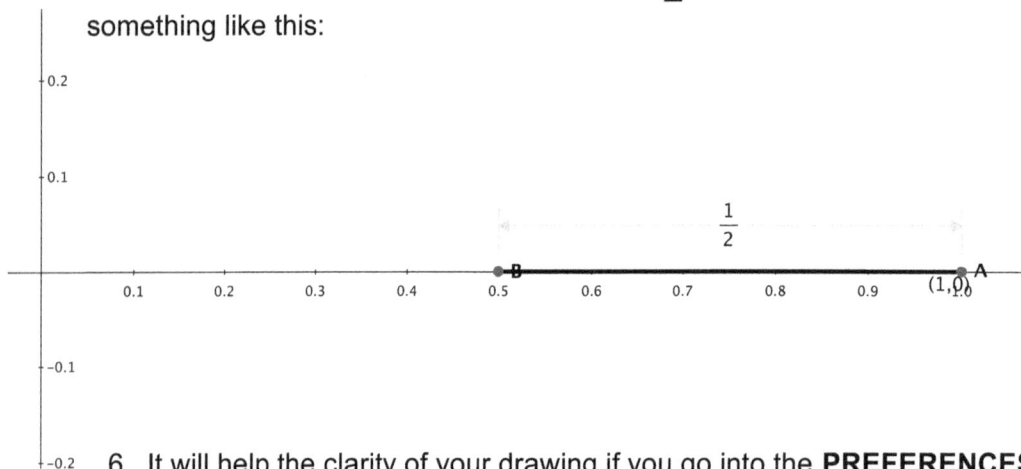

6. It will help the clarity of your drawing if you go into the **PREFERENCES** in the **GEOMETRY** tab and set the **LINE / STYLE** to **SOLID** and the **Thickness** to **4**.

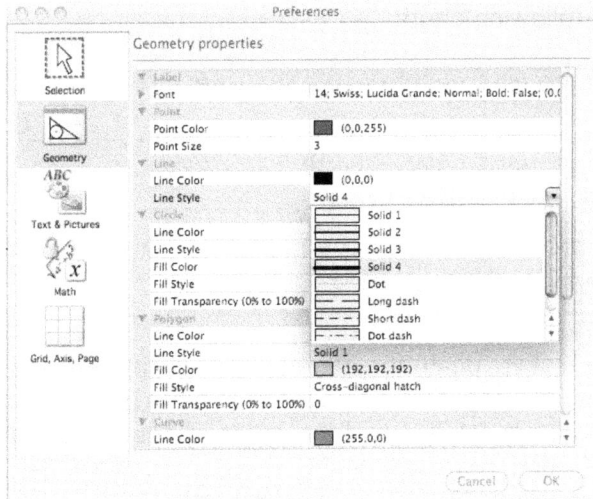

7. Identify *specifically* what each of the following represents for our frog:

A

B

Segment AB

8. We will represent the next jump by dilating segment **AB** by a factor of one-half towards the origin. Select segment **AB** and then choose **CONSTRUCT-**

DILATION ( ). Click on the origin as the point of dilation and a scale factor will appear. Change this value to ½.

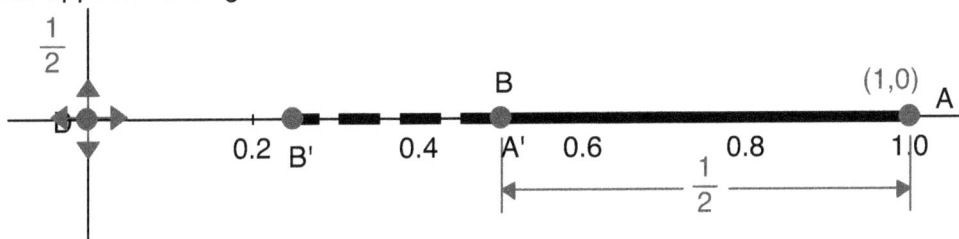

9. Measure the symbolic length of this new segment to confirm you have done the dilation correctly. Repeat the same process for *three* more jumps (construct the dilations **and** measure the segments) towards the origin. You may have to zoom in closer to the origin as the segments get very small. Have your teacher check your drawing before proceeding.

Teacher Check:_____

10. Complete the following chart:

Jump #	Distance jumped	Distance from destination
start	0	
1		
2		
3		
4		

11. Now suppose that we dilate the segments by a different scale factor, let's say 1/3. Change each of the four scale factors to 1/3 by double-clicking on each of them and editing them. Notice that the segments no longer line up and there are gaps between them.

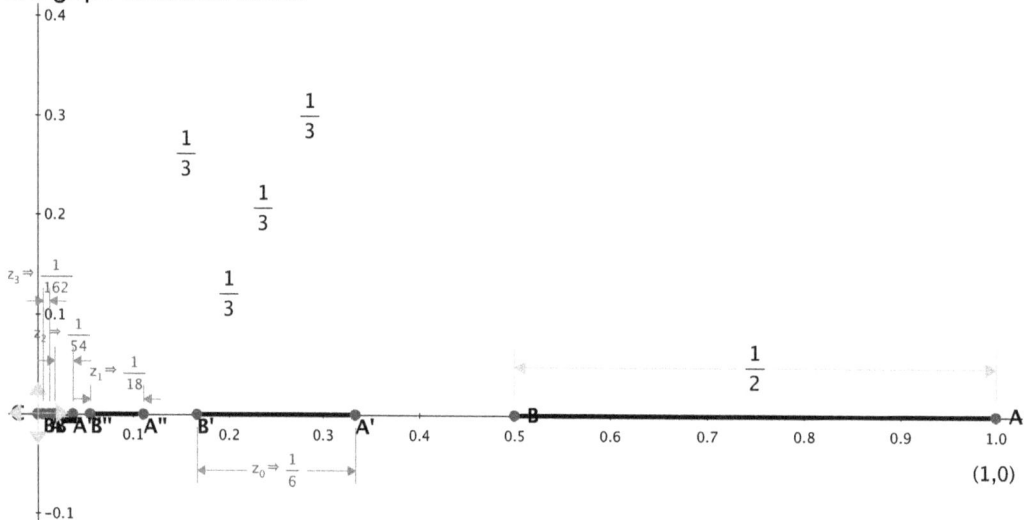

12. What is the length of the gap between points **B** and **A'**? You can use **GX** to calculate this, but you should be able to answer it yourself!

13. The next thing to consider is the length we would have to make the original segment (**AB**) have to be so that there are no gaps using this scale factor. We can use **GX** to answer this question. Double click on the length of segment **AB** and change it to **S**. Now select B and A' and measure the symbolic length between them. Record the expression below:

BA':_____

14. Since we want there to be no gap, set this expression equal to 0 and solve it. Record the value of **S** below.

S:_____

15. Verify your calculation by changing the value of **S** to your answer. It should eliminate the gaps between the segments. Have your teacher check your drawing.

Teacher Check:_____

16. Immediately undo this change (**CNTRL-Z**) so the gaps are present again. Let's generalize for *any* scale factor. Change all four of the scale factors to **r**.

17. What is the expression for the length of the gap?

18. In the space below, show all work to calculate the length of the original segment that is needed to close the gap.

19. Change this length in your drawing to verify that it is correct, save it, and have your drawing checked one more time.

Teacher Check:_____

20. Complete this chart again, this time with the generalized scale factors and segment length.

Jump #	Distance jumped	Distance from destination	Total distance traveled
start			
1			
2			
3			
4			

21. Write the series that would represent the frog jumping "infinitely". This means the frog actually reaches the destination and covers the total distance.

22. Write the equation for the infinite series that represents the total distance traveled.

23. Factor the original segment length out of this and rewrite the expression below.

24. Divide both sides of this equation by this factor.

25. This result is the formula for the sum of a geometric series that starts with an original term of one. The common ratio between two successive terms is **r**, and as long as we use a scale factor less than one, this formula will find the sum. If a geometric series begins with a different initial term, then we simply need to multiply that starting term into each term as well as the sum.

So given an initial value of **a**, the formula would become:

$$a\left(1 + r + r^2 + r^3 + r^4 + \ldots\right) = a \bullet \frac{1}{1-r}$$

or:

$$a + ar + ar^2 + ar^3 + ar^4 + \ldots = \frac{a}{1-r}$$

Achilles: So this formula helps explain that the frog really can reach me?

Tortoise: That's right! It shows that a finite quantity can arise from an infinite process. This is the foundation of the study of Calculus!

Calculus is the study of the mathematics of change and what happens as finite beings try to understand and explain the ideas of the infinite. The Greeks were dealing with these ideas thousands of years before we discovered the mathematical tools that help explain them.

Quick Questions

1. Which of the following infinite series are geometric?

a. $2 + 4 + 6 + 8 + \ldots$

b. $1 - 2 + 4 - 8 + 16 - \ldots$

c. $4 + \dfrac{4}{3} + \dfrac{4}{9} + \ldots$

d. $-2 - \dfrac{4}{5} - \dfrac{8}{25} - \ldots$

2. For the geometric series in the previous question, identify the common ratio, and use the formula to calculate the sum.

3. Using GX, construct a circle of radius one, then use dilations to represent a series of circles with radii that are 2/3 of the original circle.

    a. Write the infinite series of radii and calculate their sum.

    b. Write the infinite series of the areas and calculate their sum.

Decimal Diversion

Tortoise: We can use this same tool to explain some of the same contradictions that arise in basic computational mathematics.

Complete this chart:

Fraction	Decimal Equivalent
$\dfrac{1}{9}$	
$\dfrac{2}{9}$	
$\dfrac{3}{9} = \dfrac{1}{3}$	
$\dfrac{4}{9}$	

Many students observe this, and similar patterns with repeating decimals when they first begin decimal computations. If we follow this pattern to its logical conclusion, we arrive at the fraction $\dfrac{9}{9}$, which we know intuitively to be equal to 1. This leads us to a somewhat puzzling conclusion that Zeno most likely would have enjoyed:

$$.\overline{9} = 1$$

They are close or approximately equal- they are identical! Some students find this difficult to accept.

One argument to help illuminate this comes from the understanding of the real number system. The real numbers are an infinite set of numbers. But they are also *more* infinite than other infinite sets, such as the natural numbers. Take two consecutive natural numbers, like 1 and 2. There are an infinite number of real numbers between them. In fact, if we pick two of these numbers, say 1.2 and 1.3, there are an infinite number of real numbers just between those two. No matter how close together we make two real numbers, there are always an infinite number of real numbers between them. This is called the **density property**, and we say that the real numbers are **dense**.

If you aren't buying the argument above, try to place just one number between:

$$.\overline{9}$$
$$1$$

$$?????$$

If these numbers were not equal, you would be able to place an infinite number of numbers between them, so surely finding just one wouldn't be a problem? The reason you can't do it is because they are in fact the same number!

Achilles: Hey, are you calling *me* Shirley now?

This is actually Zeno's Paradox in another form. If you started at zero and jumped 0.9 of the distance, and then jumped 0.9 of the remaining distance, etc., what would be the total distance traveled?

This means that a number like $.\overline{9}$ can be written as the infinite geometric series:

$$.9 + .09 + .009 + ...$$

or

$$\frac{9}{10} + \frac{9}{100} + \frac{9}{1000} + ...$$

The common ratio is 0.1 and the starting value is .9, so using the formula, the sum of the series is:

$$\frac{a}{1-r} = \frac{.9}{1-.1} = \frac{.9}{.9} = 1$$

Quick Questions

Write each of the following repeating decimals as infinite geometric series and use the formula to verify the sum of each.

1. $1.\overline{6}$

2. $.\overline{03}$

3. $.1\overline{9}$

4. $.\overline{123}$

hmm... infinite geometric series... i wonder if they're edible?

Lesson Three

Purpose: Use two variables to illustrate Zeno's paradox of the tortoise and Achilles.

Tortoise: Ok big guy, it's finally your turn. Here is the original paradox of Zeno:

> *"In a race, the quickest runner can never overtake the slowest, since the pursuer must first reach the point whence the pursued started, so that the slower must always hold a lead."*

Achilles: Hey, I don't see my name in there anywhere!

Tortoise: True! We were inserted into the story later by other writers.

Achilles: I think I see where this is heading. You have a head start. Even if I am travelling faster than you (and it is MUCH, **MUCH** faster), I will have to reach the point where *you* started. In the meantime you will have covered a new distance, no matter how small it is! So I will have to reach *that* point as well. Every time I reach that new point you will have covered more ground and therefore, I cannot catch you. Is that it?

Tortoise: You've got it! Let's use a *Geometry Expressions* model to illustrate it.

Informative Instructions

1. Create a **NEW** GX drawing and save it as *lesson3.gx*

2. To represent the paradox in a uniquely mathematical way, let's turn to equations of lines and use two lines, one to represent each of our characters.

3. Turn the coordinate axes on by pressing ⊞. Graph the function (⬛) $y = x$. This line will represent Achilles. Graph another line to represent the tortoise. Enter $y = a + bx$ (Geometry Expressions does not use implicit multiplication so you actually must enter a+b*x). The tortoise has a head start (**a**) and is travelling at a slower rate than Achilles (**b**). Adjust the settings in the **Variables** tool panel so that **b** varies from -2 to 2 and set **b** to equal 0.5. **a** should be equal to 1, if not, change it. Adjust the window using the zoom tool to match the picture below. You may use the **text** tool to label the lines for the Tortoise and Achilles.

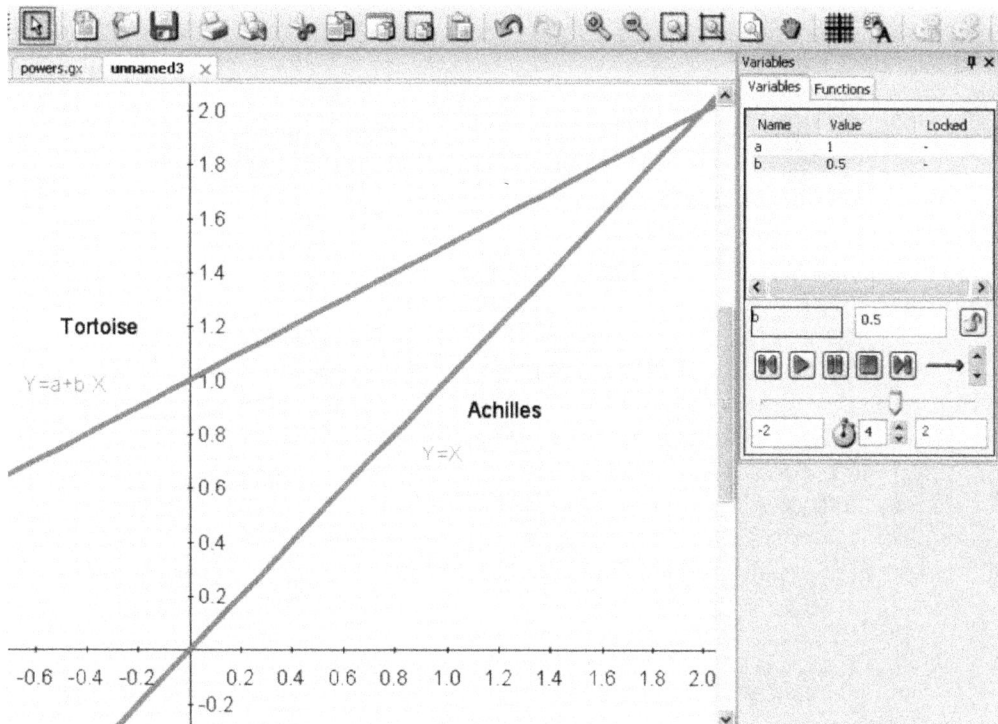

4. The tortoise is now starting one unit ahead of Achilles and Achilles is traveling twice as fast as the tortoise. You can think of the x-axis as representing some arbitrary unit of time and the y-axis as representing some distance. From the drawing, would you conclude that Achilles catches the tortoise? Explain your answer.

The paradox develops as you look at what happens for a particular distance. If the tortoise starts **a** units ahead of Achilles, for example, Achilles has to first make up that distance if he hopes to catch his tortoise friend. But while Achilles travels the first **a** units, the Tortoise will cover some additional distance. Let's show this on our diagram. In order to do so, we will need to draw parallel lines and points of intersection. Because of some potential conflicts when doing more complicated calculations, **GX** will not currently allow you to create intersection points on the functions. We can easily work around this by creating line segments that are attached to the two lines.

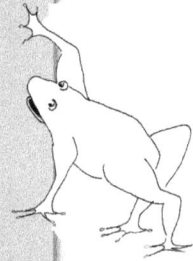

5. Using the segment tool ( ), create a segment **AB** on the Tortoise line. Make sure the point **A** is on the left of the y-axis and the point **B** is to the right of the point of intersection. Repeat this process to construct **CD** on the Achilles line as shown below.

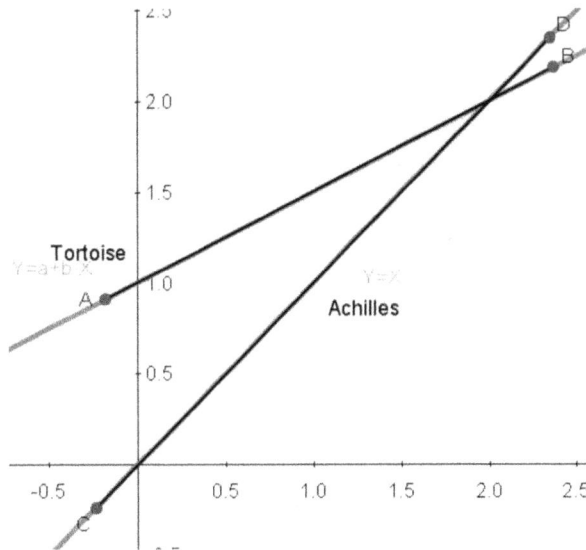

6. Now draw a segment from the intersection of the y-axis with the tortoise line and the origin (which is of course the intersection of the Achilles line with the y-axis). This represents the distance Achilles must cover to reach the point from which the Tortoise departed.

7. What is the distance Achilles must travel?

Distance:_____

8. Based on the equation we are using, what is the time interval required to cover that distance?

Time:_____

9. Select point **E** and the x-axis and choose **Construct-**

Parallel ( ). Using the point tool ( ), construct the intersection point of the parallel line and the segment on Achilles' line. Select the parallel line and choose **CNTRL-H** to hide it. Construct the segment from point **E** to this new point **G** and measure its length to verify your answer in the previous question (use a **SYMBOLIC** measurement).

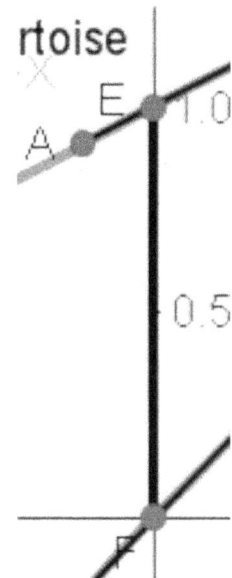

10. However, during that same period of time, the tortoise will have also been traveling. In terms of a and b, how far will the tortoise have traveled?

Tortoise distance:_____

11. To verify this, select point **G** and the x-axis and **Construct-Perpendicular** (

⊿). Construct the intersection point (**H**) of this perpendicular with the Tortoise's line, and hide the perpendicular. Create and measure the symbolic length of the segment between these two points. Have your teacher check your drawing.

Teacher Check:_____

12. Achilles now has to cover this new distance between himself and the Tortoise. While he does, the Tortoise will continue to move at his own pace, and therefore will stay ahead of him by a decreasing, but measurable distance. This is the paradox that Achilles can never catch up to the tortoise! Repeat steps 9-11 for the distance from **G** to **H** and answer the questions below.

Distance covered by Achilles:_____

Time for Achilles to cover that distance:_____

Distance covered by Tortoise in that time:_____

13. Repeat this process for one more unit of time. The following diagram has hidden some of the distracting elements. Label each distance that has a question mark next to it.

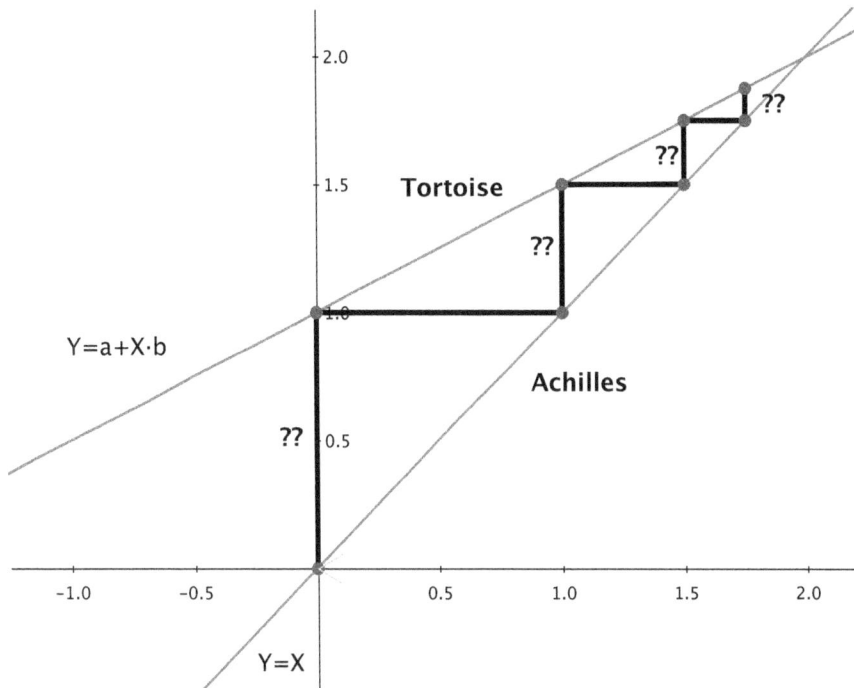

14. The drawing above represents the distance traveled after 4 units of time. We could represent the total distance traveled using a series of 4 terms. Write this series below:

Total Distance (4 time periods):_____

15. What is the next term of the series? If you cannot find the pattern yet, repeat the process above to draw the next time period and use **GX** to measure it.

Next term:_____

16. You should recognize this as a geometric series. Identify the starting term and the common ratio.

Starting term:_____   **Common Ratio:**_____

17. Use the formula developed in the previous lesson to find the infinite sum of this series.

Sum:_____

18. This represents the place at which Achilles *"really"* will catch the Tortoise, or the intersection point on the graph. As a nice confirmation of this, construct the

point of intersection between the two lines and choose to calculate the coordinates of this point. It should match your results in the previous question and is a nice visual representation of the formula for the sum of an infinite series.

Quick Questions

To answer the following questions, you need to clean up your drawing from the previous section by hiding unnecessary objects and labels and perform a few more iterations of the process. If you don't have the time or patience for that, you can obtain the file *cleanozeno.gx* from your teacher or the geometry expressions website.

1. Using the slider from the variables tool panel, adjust the value of **b.**

2. What happens to the intersection point as the value of **b** increases from .5 to 1?

3. When you make this change, explain how the situation between Achilles and the tortoise is changing?

4. Set **b** to exactly one. What is true about the drawing?

5. What does this change mean about the Tortoise?

6. If **b** is greater than one, what does this mean about the Tortoise and Achilles' prospects of catching it?

7. If a fellow student wanted to use the formula to find the sum of an infinite series in which the common ratio between two successive terms was greater

than one, what would you say to them and how would you help them to understand your answer?

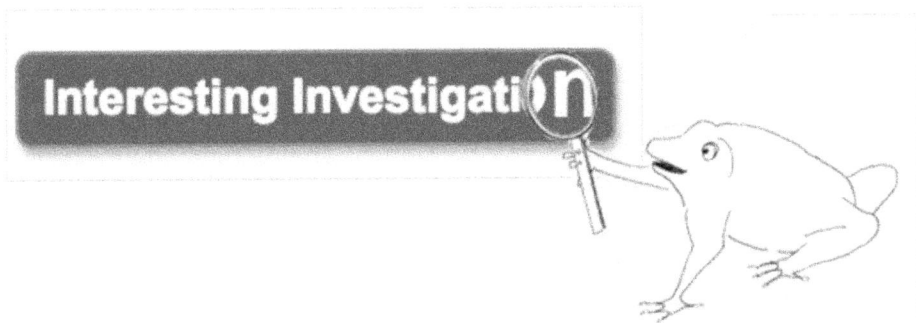

Interesting Investigation

Take the value of **b** and make it negative. We are not sure there is any relevant parallel to Achilles and the Tortoise, but it sure is an interesting looking result and if you have studied fractals and Euler's method, this might look eerily familiar. Discuss with your partner any possible meaning to this situation. Try the same idea making **b** less than -1.

Lesson Four

Purpose: Introduce a similar technique to Archimedes method using an absolute value function. Gain some practice with areas and basic geometry using Geometry Expressions.

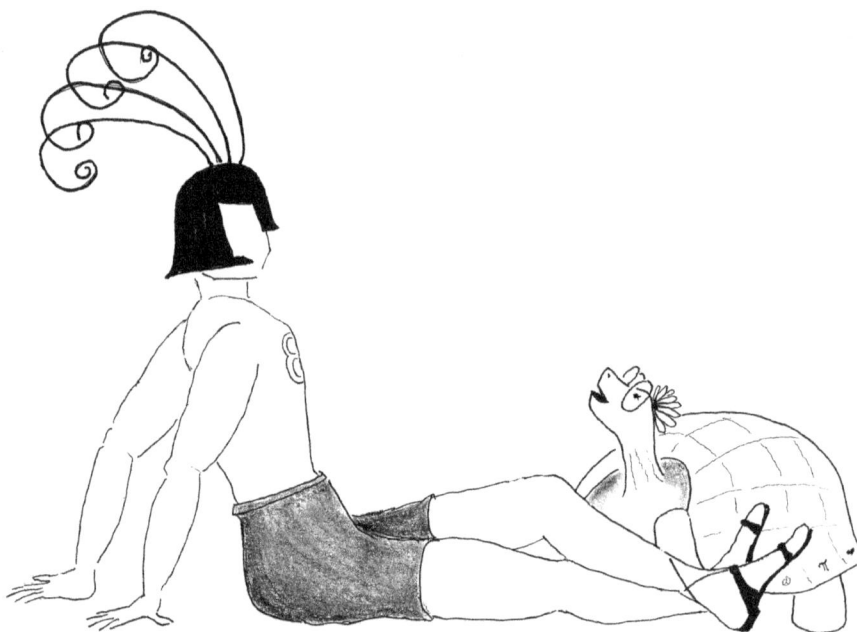

Tortoise: Several hundred years after Zeno, another famous Greek scholar, Archimedes, came closest to pulling all of the ideas of the infinite together before the discovery of the tools of modern calculus. What has become known as *Archimedes' Method* is a technique for finding the areas of segments of a parabola called **quadrature**.

Achilles: And this is related to the geometric series we've been tackling?

Tortoise: Yes! First, let's introduce some of the basics and the Geometry Expressions tools we need by studying a simpler example using an absolute value function.

Informative Instructions

1. Create a **NEW** GX drawing and save it as *lesson4.gx*

2. Turn the coordinate axes on by pressing ▦. Press it several times. You should see that this is a toggle key that alternates between no axes, axes, and axes with grids. Return it to the axes setting.

3. Press ◺ to draw a function. Type the equation Y=|X| in the dialogue box as follows:

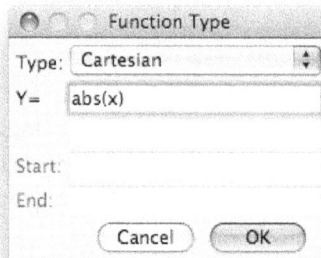

Function Type	
Type:	Cartesian ⇕
Y=	abs(x)
Start:	
End:	
	Cancel OK

4. Place a point (▱) on the function somewhere in Quadrant II. We need to assign it a value that will change along with the value of the function. Press **ESC** or ▨ to deselect the point tool then **shift-click** on both the point and the function to select both of them. Press **CONSTRAIN-POINT PROPORTIONAL (**

⬚). **GX** will suggest the parameter **t** for this value. Change it to **x**.

5. Select the point and choose ⬚ to calculate the **SYMBOLIC** coordinates of point **A**.

6. Because the algebra of absolute values can sometimes be tricky, we need to adjust a setting to simplify some of the calculations. Open your **PREFERENCES** for **GX** and choose the **MATH** tab. In the **OUTPUT** section, change the **USE**

ASSUMPTIONS setting to **TRUE**. This will allow **GX** to interpret the absolute values as they are drawn on the screen which will, in turn, make some of the Algebra a bit easier for us to handle.

7. If your original setting was **FALSE** before you changed the Assumptions, your coordinates will remain unchanged. Either delete the output and recalculate, or change the **OUTPUT PROPERTIES – Use Assumptions** for this selected output with the right-click context menu.

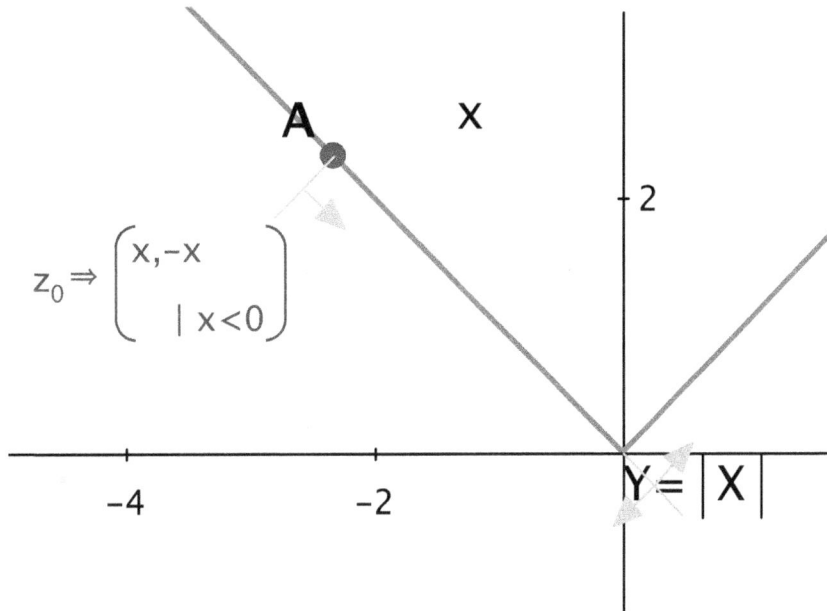

$$z_0 \Rightarrow \begin{pmatrix} x, -x \\ | \, x < 0 \end{pmatrix}$$

$$Y = |X|$$

This change reflects that **GX** is assuming the coordinates of **A** based on its position in the drawing. In this example, the x coordinate is assumed to be negative ($|x<0$), which means the y coordinate would be positive, or the opposite of x.

8. Place a second point on the function somewhere in Quadrant I. We want this point to be translated an arbitrary distance away from point A (in this case we are calling that distance **h**), so constrain the point B proportionally by **x+h**.

9. Follow the same procedure to place a third point on the function and, this time, constrain it to **0** so it is fixed at the origin.

10. Using the polygon tool, create triangle **ABC**. Select and measure the **SYMBOLIC AREA** of the interior of the triangle.

11. At this point, your model should look like this:

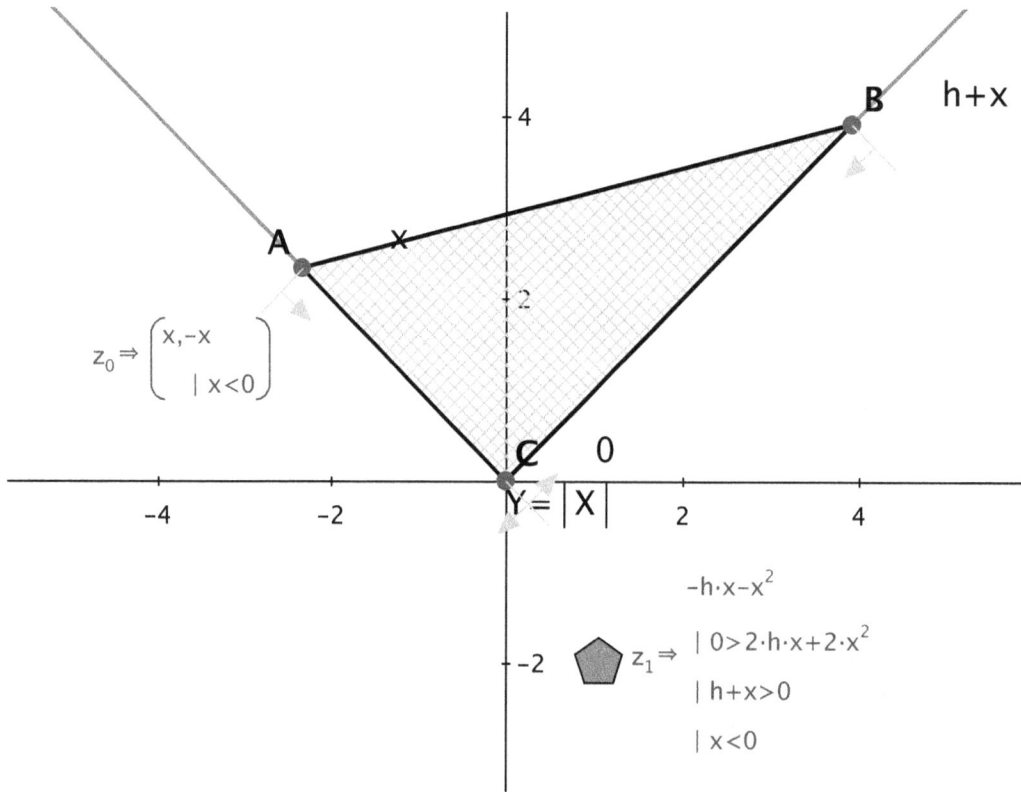

$$z_0 \Rightarrow \begin{pmatrix} x, -x \\ | \; x < 0 \end{pmatrix}$$

$$z_1 \Rightarrow \begin{array}{l} -h \cdot x - x^2 \\ | \; 0 > 2 \cdot h \cdot x + 2 \cdot x^2 \\ | \; h + x > 0 \\ | \; x < 0 \end{array}$$

$Y = |X|$

$h + x$

We want to verify that this area is correct using some geometry, so it is important to be sure your model is correct before we proceed. Have your teacher check it.

Teacher Check:_____

To find the area of the triangle **ABC**, we will consider it as two triangles with a common base, **AD**. You should be able to determine the altitudes and bases of each triangle

1. We want to place a new point on the absolute value function between **B** and **C**. In order to do so, place a point *to the right of* **B** on the function so it is not on the triangle. Constrain it proportionally to be **–x**, then adjust it to lie between **B** and **C**.

2. Construct the segment **AD**.

3. Place a point on the intersection of **AD** and the **y-axis**.

Quick Questions

1. Explain why we constrained **D** to be –x instead of x.

2. Put aside **GX** for a moment and use your geometric skills to identify the lengths of the indicated segments in the drawing below:

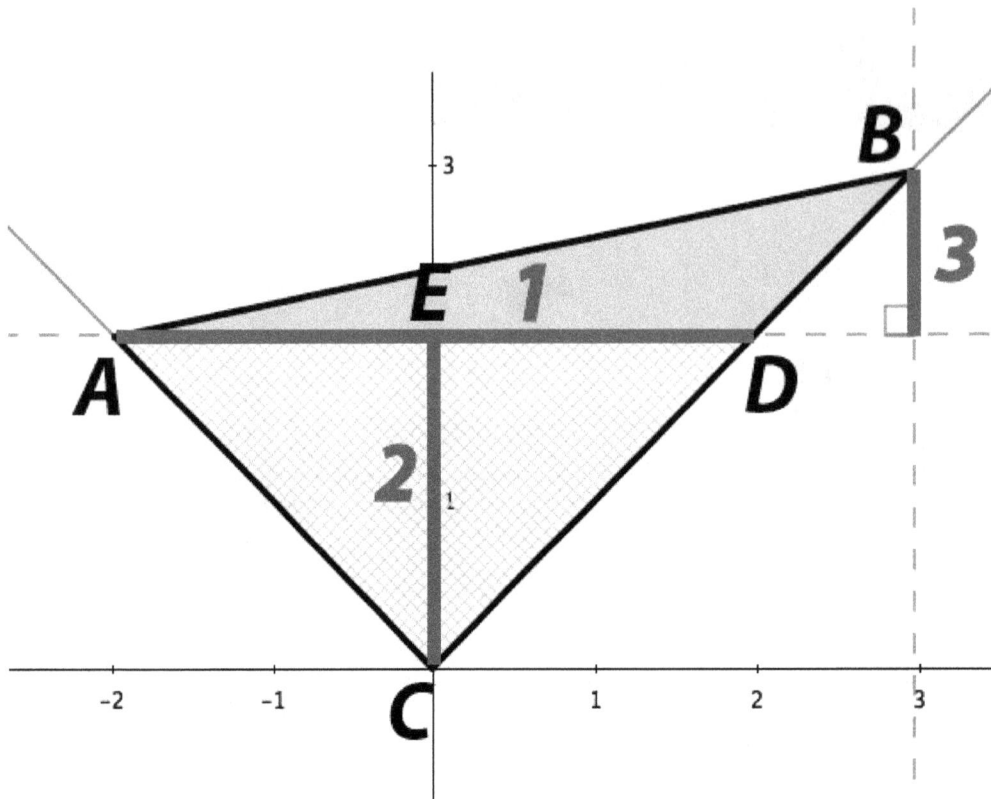

1=_____   2=_____   3=_____

3. Show all work in the space below to calculate the indicated areas.

Triangle ACD   **Triangle ABD**

AREA ACD: **AREA ABD:**

4. What is the total area of triangle **ABC**?

5. Verify your results with the answer you obtained previously.

44

6. **Save** your file and have your teacher check your final diagram.

   Teacher Check:_____

Lesson Five

Purpose: Approximate the area between a parabola and a line using Archimedes' quadrature method by addition of areas.

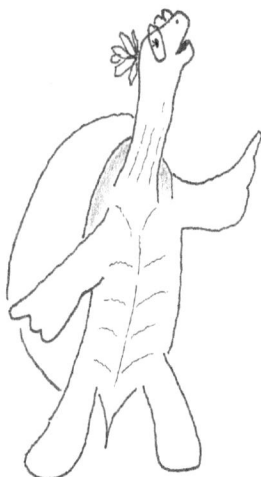

Tortoise: Archimedes is considered one of the greatest mathematicians who ever lived. One of his papers is called "Quadrature of the Parabola".

Achilles: The *Who*, of the *what*?!

Tortoise: Don't be such a heel.

Achilles: Hey, don't go there! Let's get on with it.

Tortoise: Okay, Okay. Archimedes was thousands of years before his time in terms of foreshadowing methods that weren't truly sorted out until the 17th century. But Archimedes didn't have the coordinate system to describe these relationships with equations. So he approached the problem from a purely geometric point of view. In his paper, he investigated the area of a parabolic segment, the finite portion of a parabola between two points on the curve. He showed that a particular triangle inside that parabolic segment has exactly four-thirds of the area of the entire segment.

Achilles: Why don't you just show me!

Tortoise: Sure! Fortunately, we have **GX** to help!

Informative Instructions

1. Create a **NEW** GX drawing, turn on the coordinate axes, and save it as **_lesson5.gx_**

2. Draw a new function and type in the equation $y = x^2$

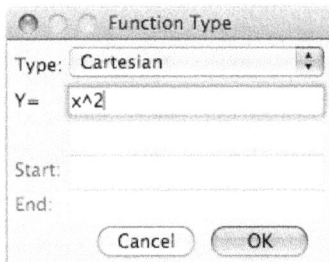

3. Just as we did in the previous lesson, place two points on the function and constrain them proportionally to be **x** and **x+h** respectively. Create the segment **AB**. Make sure that your **PREFERENCES** are set to **USE ASSUMPTIONS TRUE** as we did in the last lesson.

4. Archimedes was interested in finding the area of the region between the segment and the parabola as shown in the drawing below.

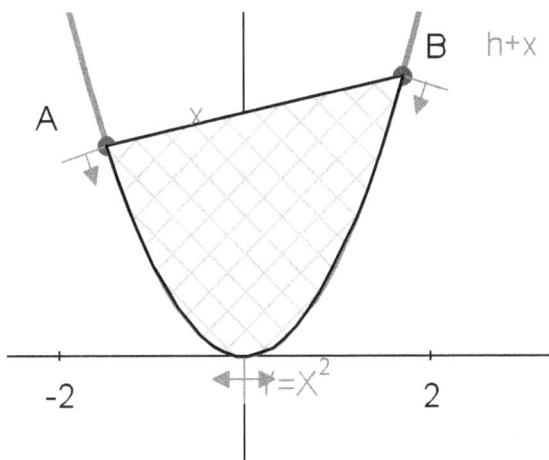

At first, a reasonable estimate of the area between the segment and the parabola would be to approximate it with a triangle. Create a third point on the parabola, between **A** and **B**. Archimedes placed this point halfway along the parabola. If point **B** is **h** units to the right of point **A**, how far to the right of point **A** should **C** be placed?

Horizontal Distance from Point A to Point C:_____

5. Constrain **C** proportionally to this position.

6. Use ⬕ to draw triangle **ABC**.

Quick Questions

1. Explain how the area of the triangle compares to the actual area between the segment and the parabola.

2. Use **GX** to measure and record the area of the triangle in the space below both numerically and symbolically.

Numeric area of ABC	Symbolic area of ABC

3. Have your teacher check your model.

Teacher Check:_____

4. Before continuing on with Archimedes' method, let's verify the symbolic area in the calculation above. Construct the midpoint of **AB**. Similar to the technique we used with the absolute value function, we will divide this triangle into two triangles that share a common base. It is actually more elegant for the parabola! The common base here is **CD**.

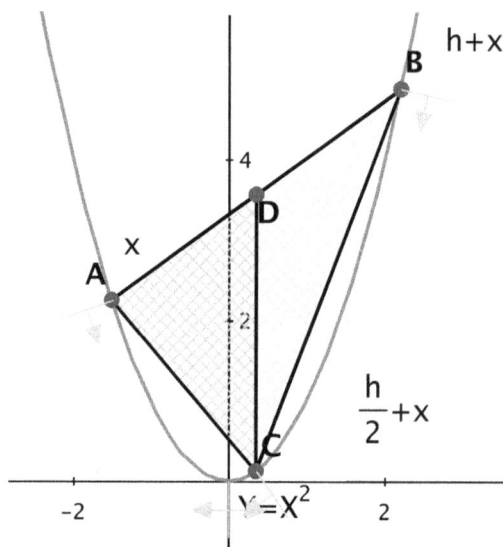

$h+x$

$\frac{h}{2}+x$

$y=x^2$

5. Identify the symbolic values of the coordinates of points A and B.

A (,   )   **B (,   )**

6. Hopefully, you didn't have to use **GX** to calculate them, so measure them now by pressing ![X] **(calculate-symbolic coordinates)**.

7. Expand the y-coordinate of point **B** in the space below.

8. One of the properties that is unique to a parabola is that the point **D** has coordinates that are equal to the average of the coordinates of points **A** and **B.** In the space below show algebraically the calculation of these average values and then verify the results using GX.

9. Use the diagram below to complete the chart and find the total area of triangle **ABC**.

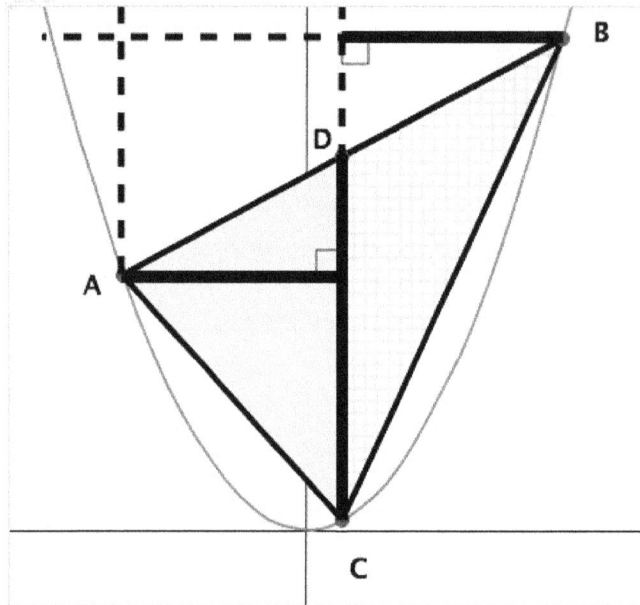

	Triangle ADC	Triangle BDC
Height		
Base		
Area		

TOTAL AREA **ABC**:_____

Your answer above should match the answer from #2. If it doesn't, don't just change it and move on! Try it again, or get some help!!

10. There are also two very important observations to make about these areas before we continue. Discuss it with your partner for a bit and then verify you have the correct observations before explaining them below as they are critically important to continuing with Archimedes' method.

       Observation #1:

       Observation #2:

Archimedes, again

1. In the first part, we approximated the area above the parabola by constructing a triangle using a midpoint on the parabola. Let's now repeat that process on a smaller scale. Construct two points on the parabola, one between A and C and the other between B and C.

2. What should the x-coordinates of each of these points be if they are midway between each of the previous points?

3. Constrain them proportionally to these halfway points.

4. Construct and measure the areas of each of the new, smaller triangles. First of all, notice that we do a MUCH better job approximating the area by adding in these two smaller triangles. There is much less white space! If we were to continue to do this over and over again on a smaller and smaller scale, we would get an approximation that approaches the actual area.

5. You may want to use different color shading for these triangles to make your drawing look more appealing. You can do this by shift-right-clicking on the small triangle interiors and changing the **Fill Properties** in the context menu.

6. Calculate the symbolic area of each triangle. Record the areas below.

       AREA of triangle **ACF**:   AREA of triangle **BCE**:

7. It should not surprise you from the algebraic exercise in the previous section that these areas are same. What is the *total* additional area we have created?

TOTAL NEW AREA:_____

8. What is the ratio of the area of the new triangles to the area of the original triangle?

Ratio :_____

9. Think of why this is true. Look at the new triangle formed in the 1st quadrant. If we were to repeat the area calculation like we did in the first example, we would use a common base formed by the mid point of the segment **BC**. Construct this midpoint.

10. If you constructed some of your points in different order, this could get a bit confusing, so use the drawing below (which should be almost identical to yours).

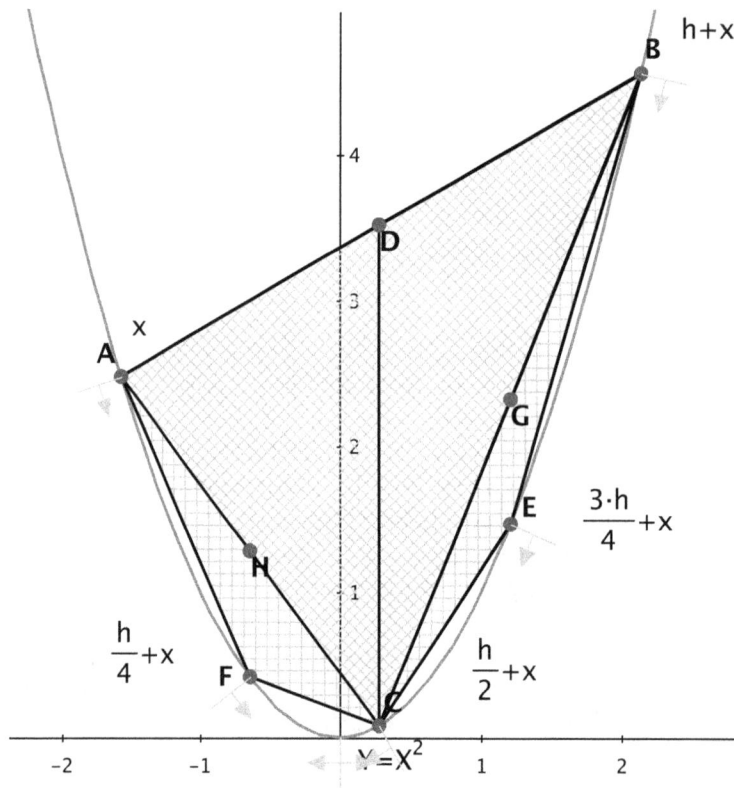

Construct **GE** and have **GX** measure the symbolic length to verify your conclusion above.

11. If we were to follow the same procedure by constructing 4 new triangles to get an even more accurate measurement, the ratio in question **8** should be true for the next iteration. Using **GX**, construct this next iteration and measure the area of the total new area added to verify your result.

AREA OF EACH NEW TRIANGLE:

TOTAL AREA OF 4 NEW TRIANGLES:

4 NEW   :
_____
2 PREVIOUS

Have your teacher check your model.

Teacher Check:_____

Achilles: I am exhausted already, and we have only done it twice!!

Tortoise: You're right! Even though we could zoom in, it's getting rather crowded on our drawing! I think you have enough of an idea to understand what would happen *if* we did it infinitely.

12. List the infinite sequence of areas below.

13. Show all work to calculate the total area for the infinite series of triangle sums.

Actual Archimedes

Archimedes original quadrature problem was actually a bit different. You still create a triangle between three points along the parabola. The middle point needs to be placed so the tangent line to the parabola through that point is parallel to the side of the triangle through the other two points. Archimedes showed that the area of this triangle is 3/4 of the area of the parabolic segment. Construct this drawing using **GX** to verify that this is true.

Teacher Check:_____

Lesson Six

Purpose: Approximate the area between a parabola and a line using Archimedes' quadrature construction by subtraction of areas.

Tortoise: In the last lesson, we showed how Archimedes found the area between the segment and the parabola by using triangles that were smaller than the actual area, and gradually adding more and more of them to get closer and closer.

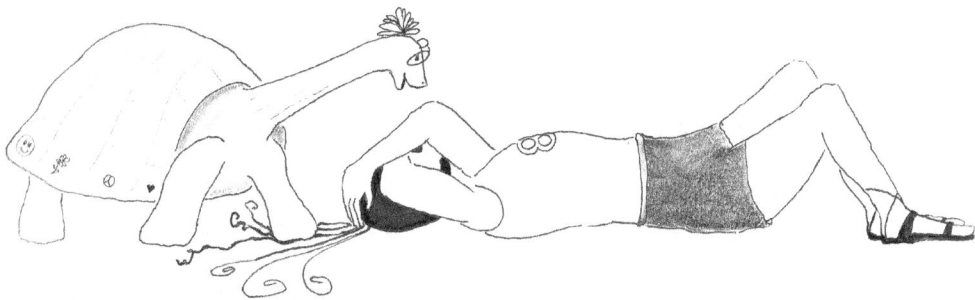

Achilles: I remember- it was exhausting!

Tortoise: Funny you should say that, it is actually called the **method of exhaustion**. It was really very clever and helped mathematicians centuries later develop the techniques of modern calculus.

Achilles: Glad I could help…so what's next?

Tortoise: Well, *you* didn't really have anything to do with it, but it's possible to repeat the same process, only using a triangle that is *bigger* than the area in question. We will take that overestimate, and subtract successively smaller triangles to approach the area from above.

Achilles: All right, let's do it!!

Informative Instructions

1. Create a **NEW** GX drawing, turn on the coordinate axes, and save it as **lesson6.gx**

2. Draw a new function and type in the equation $y = x^2$

3. Just as we did in the last few lessons, place two points on the function and constrain them proportionally to be **x** and **x+h** respectively. Create the segment **AB**. Make sure that your **PREFERENCES** are set to **USE ASSUMPTIONS TRUE**.

4. In the last lesson, we used triangles to approximate the area. Our approximation was less than the total area and we gradually approached the actual area from below. An alternate approach is to use a triangle that is larger than the desired area and subtract smaller triangles to approach it from above. Construct lines that are tangent to each of points **A** and **B** by selecting one of the points and the parabola and choosing ✂ (**CONSTRUCT-TANGENT**). Repeat for the other point.

5. Construct the intersection point **C** of these two tangents and then create the triangle **ABC**.

6. It should be fairly obvious that this triangle *hugely* overestimates the area inside the parabola. Measure the symbolic area of **ABC**.

        AREA OF **ABC**:_____

7. We will now start removing some of the extra area. Place another point on the parabola and constrain it proportionally to be halfway between **A** and **B** as we did before. Construct the tangent to this new point D and then construct the intersection points on the two previous tangents.

Your drawing should look like this:

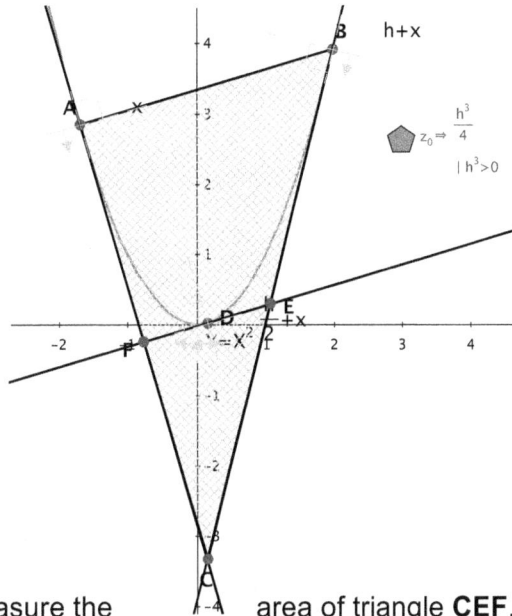

8. Construct and measure the area of triangle **CEF**.

   AREA of CEF:_____

9. What is the ratio of the area of **CEF** to the area of **ABC**?

   RATIO:_____

10. If we remove the area of **CEF** from **ABC**, we should have an improved estimation of the area in the parabola. What is the expression for the area of the resulting trapezoid?

11. Repeating this process should continue to refine our estimate. Complete another iteration by constructing two additional tangents at the quarter marks and the corresponding triangles. With the exception of some of the labels not matching, your drawing should look something like this:

12. Answer the following questions.

    a. What is the area of triangle **IJF** as labeled in the drawing above (it may be different in your drawing)?

    b. What is the **total** area of **IJF** and **KEL**?

    c. What is the ratio of this total area (**IJF + KEL**) to the area of triangle **CEF**?

    d. What is the resulting approximation of the parabola area?

    e. If we were to complete the next iteration, how many triangles would be subtracted (on just that iteration, not the total)?

f. What would the area of one of these triangles be?

     g. What is the area of the total of all the triangles constructed for this iteration?

     h. Zoom in on a section of your drawing and construct just one of these triangles. Verify your answer to part f. using **GX**. Have your teacher check your drawing.

Teacher Check:_____

i. Write the geometric series that models this process.

     j. Use the formula to calculate the sum of the series. Show all work in the space below.

This is not as straightforward as it seems. You must adjust the series to become a sum in order to use the formula!!

Lesson Seven

Purpose: Approximate the area between a parabola and the x-axis (the area below the curve. Use the arc and area calculation features of **GX** to find these areas automatically.

Tortoise: Long after Archimedes, the techniques of modern calculus made it possible to calculate the areas underneath a curve.

Achilles: Why would we want to do that?

Tortoise: You'd be surprised at the number of problems we can solve with that information. For now, let's apply the same techniques to finding these areas.

Informative Instructions

1. Create a **NEW** GX drawing, turn on the coordinate axes, and save it as **lesson7.gx**

2. Draw a new function and type in the equation $y = x^2$

3. We seem to be doing this every lesson, but place two points on the function and constrain them proportionally to the curve. This time use **t** and **t+h**. Create the segment **AB**. Make sure that your **PREFERENCES** are set to **USE ASSUMPTIONS TRUE** as we have done previously.

4. In this lesson we are interested in finding the area *under* the curve between **A** and **B**. This is coincidently something that you will be interested in when you study calculus.

5. Construct the perpendiculars to the x-axis through points **A** and **B**. Create the intersection points of the perpendiculars and the x-axis.

6. Construct the trapezoid **ABCD**, then hide the perpendiculars. To find the area under the curve, we can take the area of this trapezoid and subtract the area above the parabola.

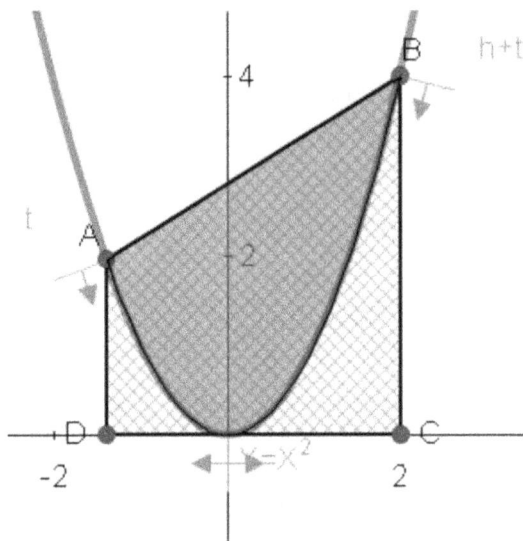

7. What is the formula for the area of a trapezoid?

8. Use the drawing above to fill out the table below with the algebraic expression for each object. You may have to look back at the previous lesson to remember the area calculation above the parabola.

Object	Length/Area
BC (one base of trapezoid)	
AD (other base of trapezoid)	
DC (height of trapezoid)	
Area of trapezoid	
Area between AB and parabola	
Area below parabola	

Precise Practice

1. Using your formula, what would the area under the parabola be from -2 to 3? Show your work in the space below.

2. Using your formula, what would the area under the parabola be from 0 to 5?

3. If we were to continue to keep the interval the same width (5 units), but continued to shift the interval up the parabola, what will happen to the area under the curve?

4. Now imagine that the interval started at 0, but was a different width from the interval in questions 1 and 2. Where would the interval need to end in order for the area to be the same as that of question 1?

Exciting Extension

Long after the Greek mathematicians used algebraic and geometric techniques to look at these areas, the methods of modern calculus were developed to simplify them. The calculations you just performed are called **definite integrals.** Integration is a tool that allows us to find the area under a curve and a definite integral is used to find this area over a specific interval. You could use a graphing calculator or CAS system to find these integrals. You will learn exactly how to do this when you take your first calculus course, just remember not to tell your calculus teacher we showed you this!

Many graphing calculators will calculate a definite integral for you. The image below shows what it would look like in Ti-Nspire using an integral, and using the formula.

Achilles: Whoa! What is that curvy thing??

Tortoise: We won't explain the integral notation here, but you most likely can figure much of it out… and we need to save SOMETHING for your calculus teacher!

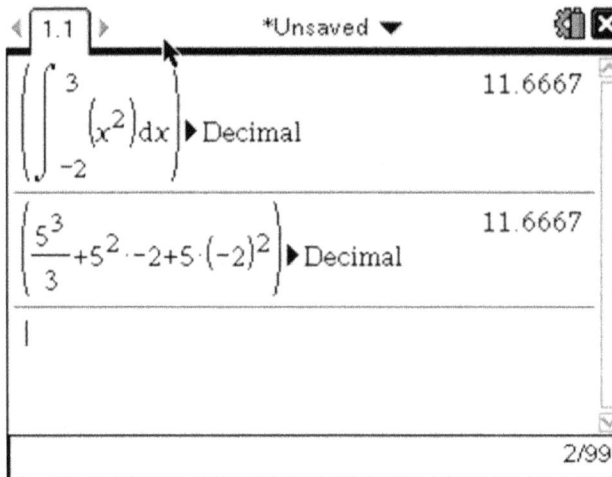

1.1 ► *Unsaved ▼

$$\left(\int_{-2}^{3}\left(x^2\right)dx\right) \blacktriangleright \text{Decimal} \qquad 11.6667$$

$$\left(\frac{5^3}{3}+5^2 \cdot -2 + 5 \cdot (-2)^2\right) \blacktriangleright \text{Decimal} \qquad 11.6667$$

2/99

1. First, use **GX** to calculate the symbolic area of the trapezoid to confirm your answer above.

2. For versions of **GX** beyond 2.2, you can choose **DRAW-ARC(**). Click on **A** and **B** to construct the arc from **A** to **B.**

3. What you have now earned the right to know through your hard work and exhaustion is that **GX** can calculate the area that we have found using

Archimedes' method!!!! Select the arc and segment **AB,** then choose to construct the area between them. Calculate this symbolic area. It should look familiar!

Achilles: Hey, wait a minute! **WHY** did you make me go through **ALL** that work in those other lessons if I could have just done that all along??

Tortoise: Many students can recall times in their math classes when the teacher first taught them how to do a problem and then showed a much easier way to solve it. Their reaction is very similar to yours. The answer is that you wouldn't have understood what is really going on if I had showed you that "shortcut" first. You most likely wouldn't remember how to do it for very long. You have now earned the right to do things the quick way!

Achilles: Well, I guess that's ok. But it does seem a little devious!

4. Back in **GX**, create an expression to calculate the area of the trapezoid minus the second area.

5. Write the value of this expression:

6. Have your teacher check your drawing.

        Teacher Check:_____

7. Now imagine if we were to change the points **A** and **B** so that **A** started at **0**. What would the value of **B**'s constraint become?

8. Substitute **0** in place of **t** in the formula we developed above. What does the area formula become?

9. If you make those changes in your drawing (change t to **0**), you should notice that the trapezoid becomes a triangle and the area formula matches the one in question 8.

Lesson Eight

Purpose: Approximate the area under a square root function.

Tortoise: Let's take all the knowledge we've accumulated over these lessons and apply it to a different function. Because it is related to a parabola, the natural place to go next is a square root function.

Achilles: Oooh, sounds radical!

Tortoise: Very good, you are catching on!

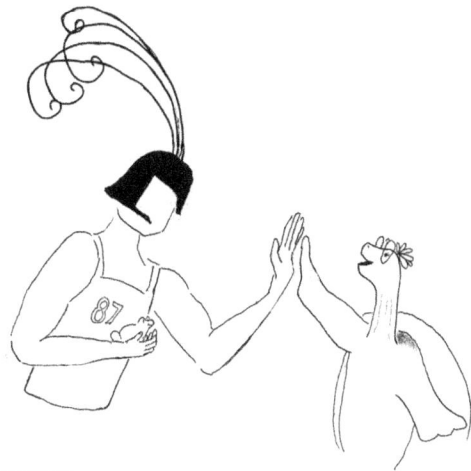

Interesting Investigation

1. Create a **NEW** GX drawing, turn on the coordinate axes, and save it as **lesson8.gx**

2. To help see the connection to our previous investigations, we need to construct another drawing of the parabola. Draw a new function and type in the equation $y = x^2$

3. Just as we did in previous lessons, place two points on the function and constrain them proportionally. To make the transition to the square root function, which is simply the same function, only reflected across the line y=x, constrain Point **A** in quadrant 2 to be **-t** and the Point **B** to be **t**. This ensures that the segment **AB** will be perpendicular to the y-axis. Construct segment **AB**.

4.

$z_1 \Rightarrow \left(t, t^2 \right)$

$z_0 \Rightarrow \dfrac{4 \cdot t^3}{3}$

$| t^3 > 0$

????

Construct the arc from **A** to **B** and then construct and measure the symbolic area between the segment and the parabola.   Record this area in the space below.

Area between segment and parabola:_____

5. Now construct the square root function $y = \sqrt{x}$. To enter a square root function in **GX**, you type **sqrt(x)**. You may already know that this function is called the **inverse** of $y = x^2$. If you switch the x and y coordinates of $y = x^2$ and then solve for y, you would have $y = \pm\sqrt{x}$. It is only a function if we consider one piece at a time, in this case the positive, or top half.

6. This parabola is a rotation/reflection of the original. The parabolic segment should have the same area. If we only measured the area between the line segment (**CF** in the drawing below), the top half of the new parabola, and the x-axis, what would be true about its measurement?

7.   The only problem is that the parameter **t** in the first area measurement is different from what it would represent on the second curve. Complete the following to discover *how* it is different.

a. Write the coordinates of point **B** (,   )

b. According to the procedure we used to find the inverse function, what are the coordinates of the point that is a reflection of point **B** across the line y=x?

(,)

c. Construct a new point, **C** on the square root function and constrain it proportionally to the curve and the x-coordinate from the previous question. To verify that it is correct, either drag point **B**, or adjust the value of **t** with the slider in the variables tool panel. You should see that point **C** moves properly in conjunction with the value of **t**.

Have your teacher check your drawing.

TEACHER CHECK:_____

d. Using the formula for the area of the original area from number 4 above, what is the area between the square root function and the x-axis from the origin to the point C?

Area under square root function:_____

8. Let's use **GX** to construct and verify the area. Place a new point on the square root function, and constrain it proportionally using 0 to fix it at the origin.

Choose [icon] to draw the arc from this new point, to point **C**.

Construct a perpendicular to the axis and create line segments to form the

boundary of the desired area. Select all three segments. Choose [icon] to construct the area, and measure the symbolic area.

9. This is actually the area in terms of the old parameter, **t**. To find the correct area in terms of the new function substitute a new parameter, let's call it **u**, in place of t^2. Replace the constraint for point **C** in your drawing with **u**. Record the resulting area in the space below.

Area under the square root function:_____

Quick Question

Explain *why* the area given in the last expression by **GX** is correct by substitution.

Lesson Nine

Purpose: To extend the area under a curve for power functions to generalize a rule for finding the areas.

Tortoise: In previous lessons we looked at the area under parabolas and square root functions. Archimedes' method only worked using parabolas, but using the capabilities of **GX**, we should be able to generalize to some other power functions.

Achilles: Sounds right up my alley... Uh, what's a power function?

Tortoise: A function of the type $y = ax^b$ is called a power function. Let's first look at the area of a triangle formed by the endpoints, and then generalize this to the area below the curve over the given interval.

Quick Question

Though slightly trivial, the simplest form of a power function is a straight line (where b=1). Look at the drawing below. What is the area of the triangle?

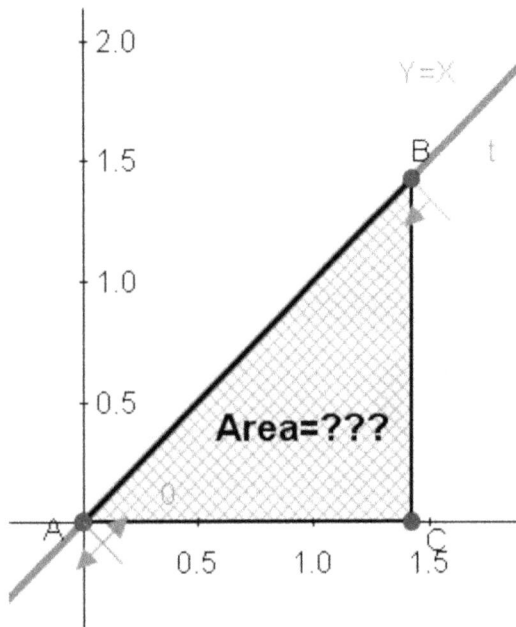

AREA=_____

Informative Instructions

1. Open a new **GX** document and save it as *lesson9.gx*. Show the coordinate grid and enter a function, $y = x^2$. Zoom in on your function. Place two points along the curve. Use $\overline{.9}$ to constrain point **A** to the origin, and point **B** to an arbitrary position along the curve, **t.**

2. Now drop a perpendicular from point **t** to the x-axis and construct the intersection point of this perpendicular with the x-axis. Use $18.12.8,\frac{16}{3}...$ to construct the triangle **ABC**. Hide the perpendicular.

3. Select the interior of this triangle and measure the area using . Record the area in the space below:

Area of triangle ABC:_____

4. Double click on the function and change it to $y = x^3$. Record the resulting area below:

Area of triangle ABC:_____

5. Do you see the pattern? Discuss this with your partner and explain the pattern in the space below. Include at least one specific area expression for a higher order power function (b=5 or higher).

6. Now let's use **GX** to calculate the areas we are really after, the area **below** a power function from **0** to **t**. First, change your function to $y = x^2$. Choose

and construct the arc on the parabola from **A** to **B**. Select the arc and

sides **AC** and **BC** and choose to construct the area under this portion of the curve. Measure it using **Calculate Symbolic Area**.

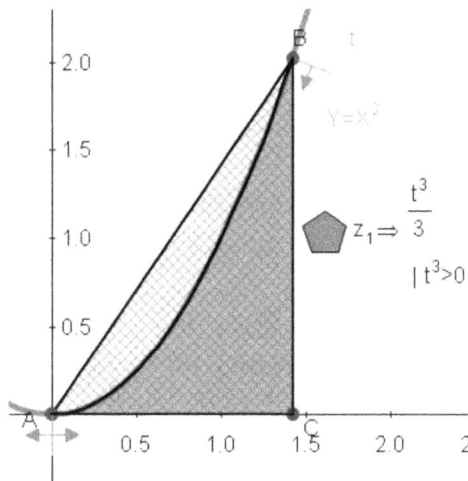

7. Change the function a couple of times until you find the pattern. Complete the table below.

Function	Area of triangle	Area under curve
$y=x$		
$y=x^2$		
$y=x^3$		
$y=x^4$		
$y=x^5$		
$y=x^6$		
$y=x^n$		

Quick Questions

1. The generalized rule for **n** is something you will investigate further when you study calculus. In the previous lesson we found the area under the square root function. Extend this rule to verify the results from that investigation.

   a. What is $y = \sqrt{x}$ expressed as a power function?

   b. Using the pattern above, what is the area under this function from **0** to **t**?

2. Let's think about specific areas for certain values of **t**. If you have changed the function to help you answer previous questions, change it back to $y = x^2$.

   Change the constraint on point **B** to 2. Observe the change in the area as you change the exponent in the power function. If you are not sure, right click on the area measurement and choose **CONVERT TO REAL**. Then experiment with changing the power function and watch what happens to the area. Describe the change in the area under the curve as the value of the exponent in the power function increases.

3. A much more fun way to get a feel for this relationship is to use the animation feature in the **Variables** tool panel. Double-click on the function to change it to $y = x^n$. If the variables tool panel is not open go to **VIEW-TOOL PANELS-Variables**. Select the variable n and change the settings to vary n from 0 to 8.

Take the slider and drag it to the left or right to observe the change in the region as the exponent changes. This should be a nice visual reinforcement of your observation in the previous question. To make it even more exciting, press the play button to animate the change. Adjust the settings so the animation plays in a continuous loop and have your teacher check the result.

Teacher Check:_____

4. Describe how your answer to question 2 would change if **t=1**?

5. Describe the change if **t<1**.

Lesson Ten

Purpose: Use a method similar to that of Archimedes to calculate/approximate the area under the normal curve and answer questions about statistically based probabilities.

In the study of statistics we are interested in the manner in which data from a particular group, called a population, is distributed. For example, if we were to measure the heights of a large group of 12th grade boys and plot them on a number line, we would most likely observe that most heights cluster around a common value. Some would be taller, and some shorter. It is also likely that one or two would be much taller or much shorter. A possible data set might look something like this:

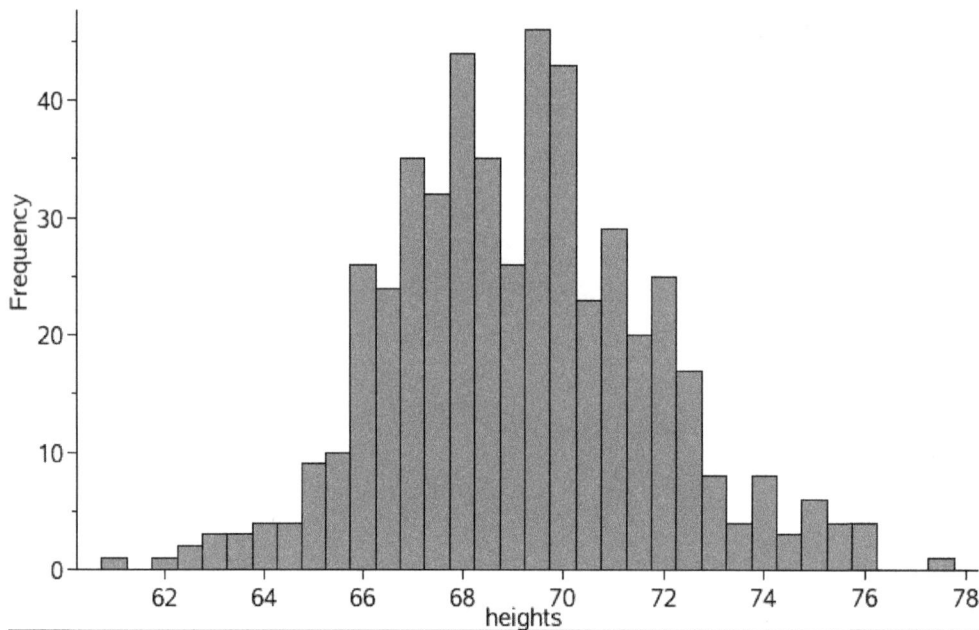

If we allow for slight variations here and there, and just look at the overall shape of this data, we might approximate it using a smooth curve like this:

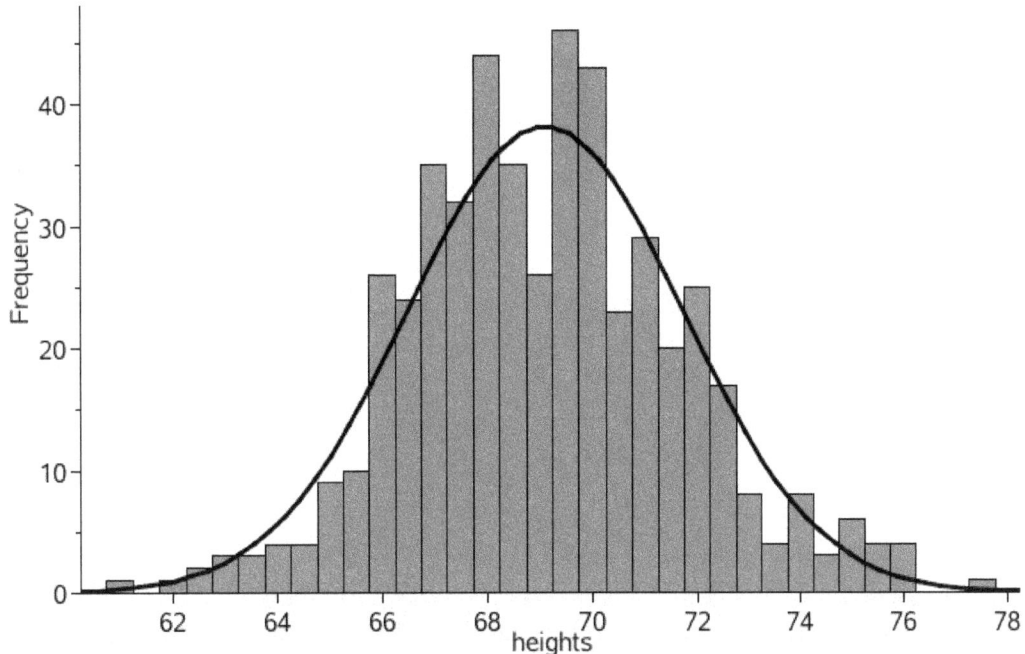

In fact, so many important situations in the real world, specifically those involving randomness, take on this shape which statisticians call a **normal curve**. Often statisticians and students studying statistics need to know the area under a specific region of this curve in order to predict the probabilities of certain outcomes. Software in which these areas have already been programmed is usually used. These type of calculations answer questions such as:

What is the probability that a randomly chosen 17 year-old boy is taller than 73 inches?

Or:

Jeff is 64 inches tall. What percentage of boys his age are shorter than he is?

Or:

We asked 1000 voters if they believed the governor is doing a good job. How likely is it that the results from our sample accurately reflect the views of ALL voters?

These area calculations are done through techniques of calculus that are usually not covered in a statistics class. But it is important that students intuitively understand them as area calculations.

Achilles: Boy, this stuff is all Greek to me!

Tortoise: The discipline of statistics is relatively new historically. While many mathematicians who studied probability knew of these ideas, it really wasn't until the last 250 years or so that we began to think in these terms. Even though Zeno or Archimedes certainly wouldn't have thought in quite the same way, it seems a natural step to apply Archimedes' ideas to estimating these areas.

Achilles: Ok, let's give it a shot!

Informative Instructions

1. Create a new **GX** drawing, turn on the coordinate axes, and save it as *lesson10a.gx*

2. When statisticians want to compare different populations that are normal (the height of a 17 year-old girl vs. the height of a 17 year-old boy) they use a **Standard Normal Curve.** This is a particular normal curve that has a center (or mean) of 0 and most of the points are within 1 unit of this mean. By converting both of the heights to a standardized value, we can compare individuals in different populations.

The equation of this curve is given by the following rather complex formula. While it is not important to dwell on the specifics of this formula, it is a rather exciting one that includes *two* of the more famous mathematical constants!

$$y = \frac{1}{\sqrt{2\pi}} e^{-\frac{1}{2}x^2}$$

If we want to generalize to any distribution, we use a normal curve that is centered at µ (the mean of the population) and has a standard deviation (the typical distance of a point from this mean) of σ.

$$y = \frac{1}{\sqrt{2\pi\sigma^2}} e^{-\frac{(x-\mu)^2}{2\sigma^2}}$$ Draw a new function and type in the equation of the

generalized normal curve. Be careful to type it in correctly! The constant **e**, raised to a power is called an exponential, so the syntax for this is **exp**. Using this notation, **e**^{x} would be written as **exp(x)**. To use σ and µ, you must use letters from your keyboard when entering the equation into the **Function** dialog, then when the graph is drawn, double click the equation and replace the letters with the Greek ones from the **SYMBOLS** tool panel. Don't panic, GX will rearrange your equation a bit. If you entered it correctly it should look like this:

3. Rescale the axes

using ⬜ (zoom to selection) to show the part of the curve that has the most visible area.

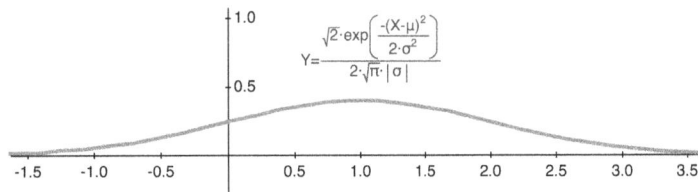

$$Y = \frac{\sqrt{2} \cdot \exp\left(\frac{-(X-\mu)^2}{2 \cdot \sigma^2}\right)}{2 \cdot \sqrt{\pi} \cdot |\sigma|}$$

4. Place a point on the curve and constrain it proportionally to the mean (**µ**).

5. Constrain a second point to be one standard deviation above the mean (**µ+σ**).

6. Experiment with changing the values of **µ** and **σ** by dragging the constrained points.

7.

Answer these two questions quickly:

1. What is the affect on the curve of changing the value of **μ**?

2. What is the affect on the curve of changing the value of **σ**?

8. Place a point on the curve an arbitrary distance (**t**) above the mean and place a second point on the curve and arbitrary distance (**s**) *below* the mean.

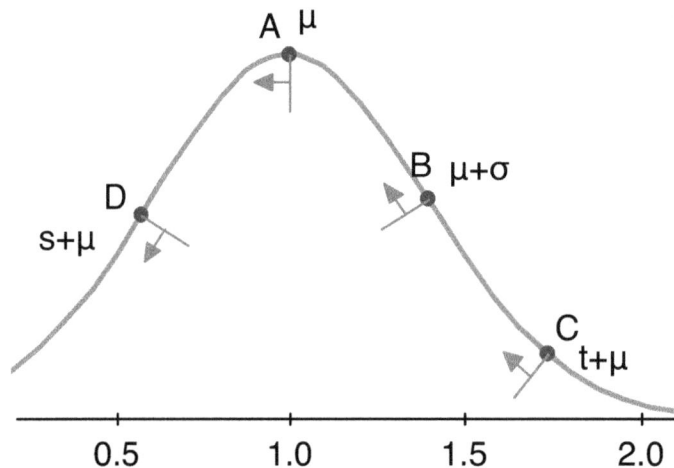

9. Drop perpendiculars to the **x-axis** from each of these points and construct the intersection point with each perpendicular and the **x-axis**.

10. After hiding the perpendicular lines, use the polygon ▱ tool to draw the trapezoid that connects these four points. Measure the symbolic area of the trapezoid.

11. While the area of this trapezoid may appear intimidating, it could be derived without much difficulty. What is important to notice is that it does not include the mean. To illustrate this, in the variables tool panel, vary the value of **μ** and it

should be fairly obvious that changing the center of the curve has no effect on the area of the trapezoid. It is simply translated left or right.

12. The standard deviation, σ, will affect the area of the trapezoid. If we make the standard deviation a constant, however, then the only values that affect the area are the position of **s** and **t**. As an alternative, let's make the distances **s** and **t** multiples of the standard deviation. Double click on the constraints and change them to be **μ+t*σ** and **μ+s*σ**. The resulting formula for the area of the trapezoid should include only the variables s and t, so we have a situation much like Archimedes and the parabola in that the area is independent of x, μ, and σ.

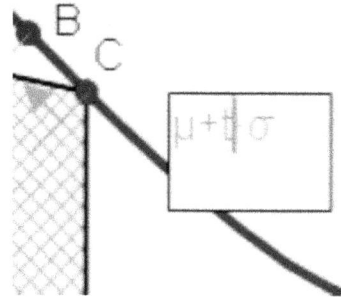

13. Save your document and have your teacher check it.

Teacher check:_____

14. In a similar fashion to the technique used in Lesson 5, we could find a closer approximation of the area by finding the point on the curve at the midpoint of these two points and using two trapezoids.

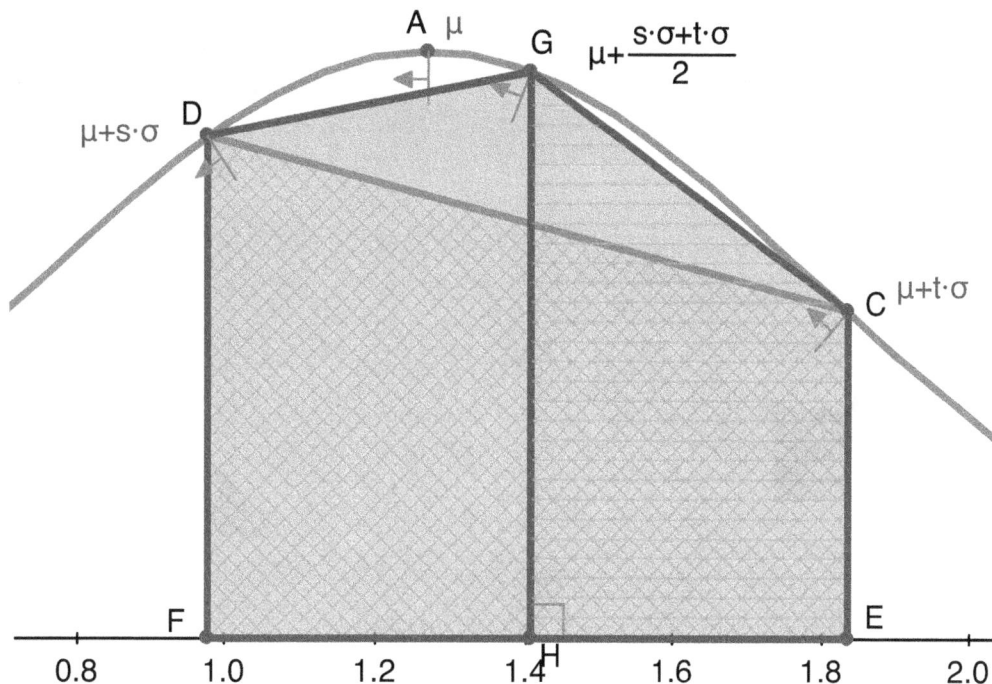

These areas would again be independent of the other variables and we could (at least in theory) write the infinite series of the sums of more and more subdivided trapezoids! In reality, the terms of the series are rather complicated and the common ratio of these terms might not be something we want to attempt to find.

Assessing Areas

Tortoise: Since we already know that **GX** will handle the area calculations, and because they would be complicated otherwise, we will skip calculating the series and go right to the answers. For our last investigation, let's use GX to establish an important concept in statistics about the normal curve.

Achilles: Wait, we're done? I was just getting into this!

Tortoise: Save your energy- we're going to race after this!

Achilles: You're on shell girl, and don't count on any paradoxes to help you out!

1. Open a new document and save it as *lesson10b.gx*.

2. Show the coordinate axes and then graph the function for the standard normal curve (mean of 0 and standard deviation of 1). In order to save some effort typing, just graph y=x (or any valid function) as the function. Return to the *lesson10a* file and select the function definition. Choose **CNTRL-C** to copy it, then return to the *lesson10b* file, double click on your newly defined function and press **CNTRL-V** to paste the normal curve definition. Delete **μ** from the equation. **Important Note:** also be sure to delete the *Y=* from the equation.

3. Rescale the axes using [zoom icon] (zoom to selection) so the x-axis goes a little more than 3 units from the origin in both directions.

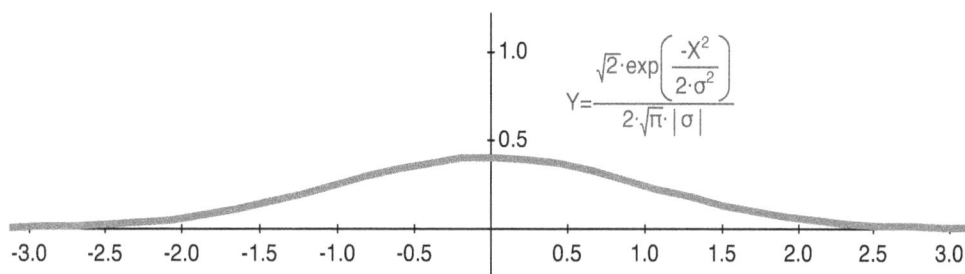

$$Y = \frac{\sqrt{2} \cdot \exp\left(\frac{-X^2}{2 \cdot \sigma^2}\right)}{2 \cdot \sqrt{\pi} \cdot |\sigma|}$$

4. Resize the curve to make it more visually appealing by changing σ to 0.5.

5. Create a point **A** and constrain it proportionally to the curve. Call this distance **z*σ**. In statistics, we refer to a **z-score** as a measure of the number of standard deviations that a value is away from the mean. By using this formula, we will insure this is exactly what z represents in this drawing.

5. Constrain a second point **B** similarly with a distance of **–z*σ**.

6. Position point **A** to the far right end of the drawing. As we move this point farther to the right, we come closer and closer to the **x-axis**. This is our infinite paradox again. The curve gets closer and closer to the **x-axis** without touching it. However, if we were to be able to extend this out to infinity, they would touch!

7. One of the most important ideas of statistics and the normal curve has to do with the percentage of data that is within a certain number of standard deviations of the mean. This is represented geometrically by the area under the curve in that region.

8. To construct this area, return point **A** to around 0.5. Once again, drop perpendiculars from **A** and **B** to the **x-axis** and construct the intersection points. Hide the perpendicular lines.

9. Use ⌐ to construct the arc from **B** to **A**. Construct the line segments that form the bottom portion of the area.

10. Select the arc, the line segments, and choose ⬠ to create the region.

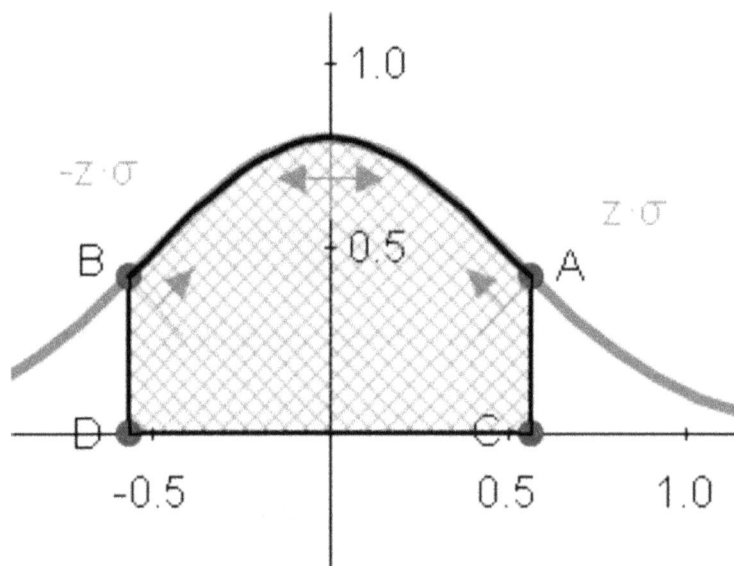

11.   Measure the real area of this region. In the variables tool panel, change **z** to 1. Record the resulting area to 3 decimal places in the space below:

Area from -1 to 1 standard deviation:_____

12. Change **z** to 2. Record the resulting area in the space below:

Area within 2 standard deviations:_____

13. Change **z** to 3. Record the resulting area in the space below:

Area within 3 standard deviations:_____

14. Write your answers to the last three questions as percentages (round the last one to the nearest tenth of a percent).

_____   _____   _____

You've just discovered what statisticians refer to as the **empirical rule**. It's a handy way to remember the approximate percentages of data within 1, 2, and 3 standard deviations of the mean.

The End

Teacher Notes

Detailed instructions are provided in the following lessons for constructing the necessary Geometry Expressions models. No experience with the software should be necessary **IF** the student begins with the first lesson and moves chronologically through the activities. However, as the lessons proceed, it is assumed the student will gain some facility with the program, and therefore fewer specific instructions and screenshots are given. The teacher may decide to skip some lessons and jump ahead to others, so it is important to understand this and avoid frustration on the part of some students learning their way around the software.

While all teachers utilize different classroom presentation styles for using technology, the author's experience reflects the set-up of these materials. They are written so the students will be basically self-sufficient with the teacher serving only as a resource and for an occasional "nudge" in the correct direction. Ideally, students should work with partners. One student should read the instructions, while the other student executes them. Students should switch roles every now and then to insure they are both using the software. Less motivated students may need help staying on task. This is the reason for including occasional "teacher checks" at certain points during the activities. Teachers may choose to utilize, ignore, or modify these depending on the level and/or motivation of their students.

Lesson One

Objectives

The student will be able to use the basic commands of Geometry Expressions.

The student will be able to recognize geometric sequences.

The student will be able to find the common ratio in a geometric sequence.

The student will recognize that the sum of certain infinite geometric series' is a finite value.

Lesson Notes

This lesson starts with some review of the basics of identifying geometric sequences and finding the common ratio in a geometric sequence. If the students are already familiar with these ideas, this can be used as review, or skipped entirely.

Exciting Examples

1.   a. The result of dividing any term by its previous term is always
$$\frac{8}{4} = 2, \frac{4}{2} = 2, \frac{2}{1} = 2.$$ So this is a geometric sequence with a common ratio of 2.

   b. This is an infinite sequence.

2.   a. not geometric b. not geometric. c. geometric, 3. d. geometric, ½

3.   a. geometric, 1.5   b. geometric, $\dfrac{2}{3}$

      c. geometric, -2   d. geometric, $-\dfrac{1}{3}$

4. $12, -8, \dfrac{16}{3}, -\dfrac{32}{9}, \dfrac{64}{27}$

5. $-81{,}920$

Quick Questions

1. Each new triangle is one half the area of the preceding triangle.

2. The area of the shaded regions will get closer and closer to zero. Use this question as a chance to introduce vocabulary like: "the area **approaches** zero". Discuss the area of the infinite triangle.

Exciting Extension

3. $\dfrac{63}{32}$

4. $\dfrac{255}{128}$

More Quick Questions

1. The area is increasing, larger at first, but smaller as we go on.

2. The total area will get very close to 2.

3.

Term/iteration number	New Shaded Area	Total Shaded Area
0	1	1
1	0.5	1.5
2	0.25	1.75
3	0.125	1.875
4	0.0625	1.9375
5	0.03125	1.96875
6	0.015625	1.984375
7	0.0078125	1.9921875
8	0.00390625	1.99609375
. . .	. . .	. . .
REALLY big number	really close to zero	really close to two
∞	0	2

4. It represents the original square that is completely shaded, or the initial term of the sequence.

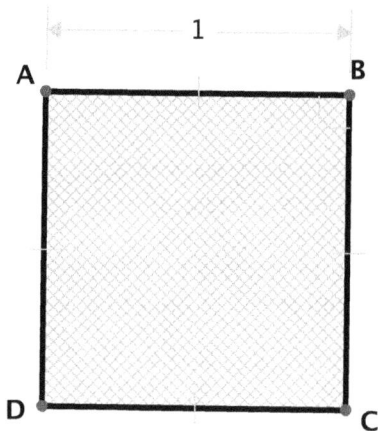

5. Model should look something like this:

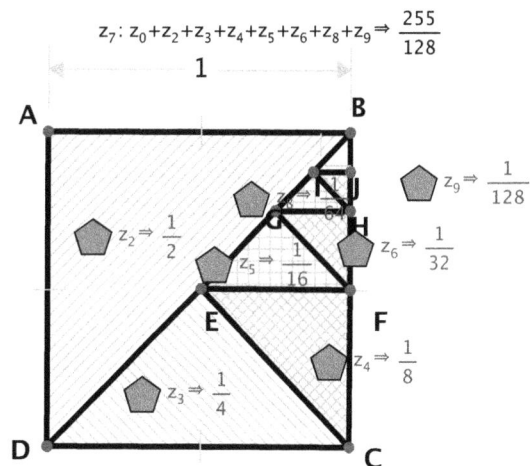

$z_0 \Rightarrow 1$

$$z_7 : z_0 + z_2 + z_3 + z_4 + z_5 + z_6 + z_8 + z_9 \Rightarrow \frac{255}{128}$$

$z_9 \Rightarrow \frac{1}{128}$

$z_2 \Rightarrow \frac{1}{2}$

$z_5 \Rightarrow \frac{1}{16}$

$z_6 \Rightarrow \frac{1}{32}$

$z_3 \Rightarrow \frac{1}{4}$

$z_4 \Rightarrow \frac{1}{8}$

Lesson Two

Objectives

The student will be able to use Geometry Expressions and Zeno's Paradox to derive the formula for the sum of an infinite geometric series.

The student will be able to use Geometry Expressions to perform dilations.

The student will be able to recognize and use connections among mathematical ideas.

The student will be able to use reasoning and methods of proof.

The student will be able to find the sum of an infinite geometric series.

Lesson Notes

In this lesson, the student uses the dilation feature of geometry expressions to discover or develop the formula for the sum of an infinite geometric series. The second part of the lesson extends these ideas to a simple, but beautiful numerical paradox involving repeating decimals that is accessible for all students and helps make a nice connection between the philosophical and quantitative aspects of infinite series.

7. **A**: The starting point for the frog
    B: The ending point for the first jump
    AB: The length of the first jump or the scale factor for the jump (because he started at 1)

9. Here is what the drawing should look like:

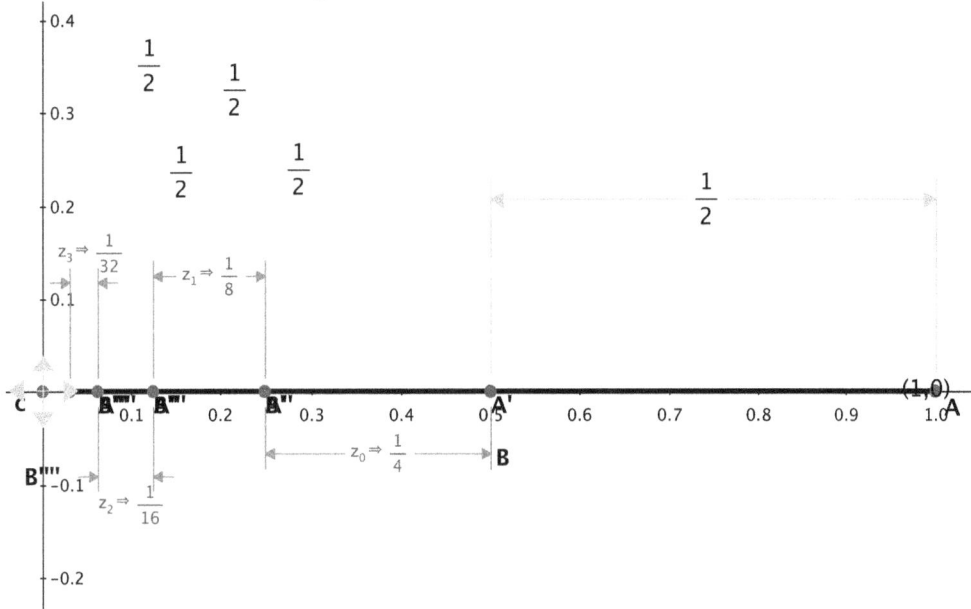

10.

Jump #	Distance jumped	Distance from destination
start	0	1
1	0.5 or 1/2	0.5
2	0.25 or 1/4	0.25
3	0.125 or 1/8	0.125
4	0.0625 or 1/16	0.0625

12. 1/6 (1/3 of ½)

13. **BA'=** $\dfrac{2}{3} - s$

14. $s = \dfrac{2}{3}$

15.

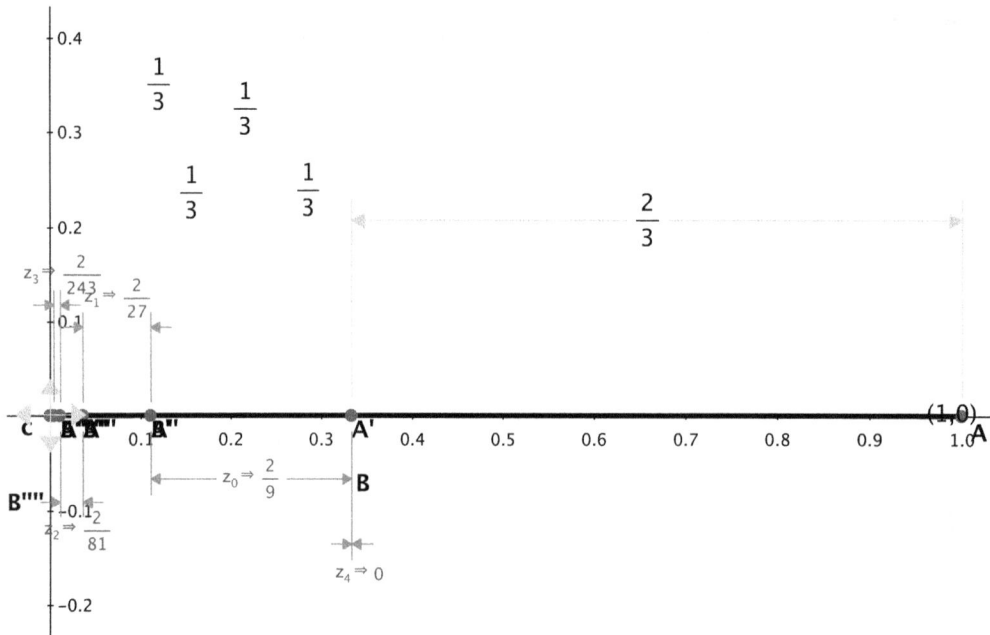

$\frac{1}{3}$ $\frac{1}{3}$

$\frac{1}{3}$ $\frac{1}{3}$ $\frac{2}{3}$

$z_3 \Rightarrow \frac{2}{243}$ $z_1 \Rightarrow \frac{2}{27}$

0.4

0.3

0.2

0.1

c A A''' A'' 0.1 A'' 0.2 0.3 A' 0.4 0.5 0.6 0.7 0.8 0.9 (1,0) 1.0 A

$z_0 \Rightarrow \frac{2}{9}$ B

B'''' −0.1 $z_2 \Rightarrow \frac{2}{81}$

$z_4 \Rightarrow 0$

−0.2

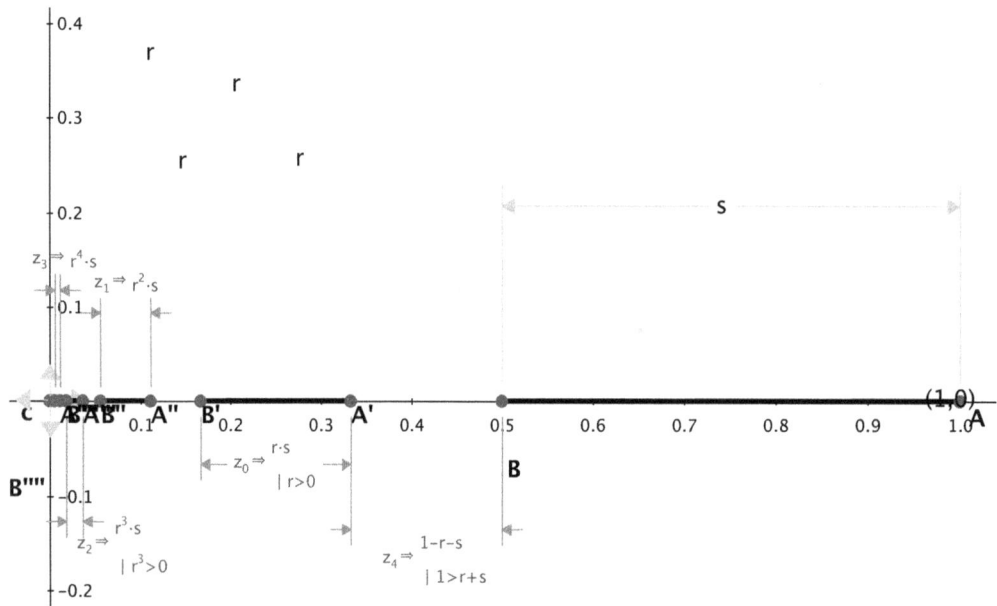

0.4

r r

r r

0.3

s

0.2

$z_3 \Rightarrow r^4 \cdot s$ $z_1 \Rightarrow r^2 \cdot s$

0.1

c A'''AB''' A'' 0.1 A'' B' 0.2 0.3 A' 0.4 0.5 0.6 0.7 0.8 0.9 (1,0) 1.0 A

$z_0 \Rightarrow r \cdot s$ $| r > 0$ B

B'''' −0.1 $z_2 \Rightarrow r^3 \cdot s$ $| r^3 > 0$

$z_4 \Rightarrow 1 - r - s$ $| 1 > r + s$

−0.2

17. $1 - r - s$

92

18.

$$1 - r - s = 0$$

$$1 - r = s$$

19.

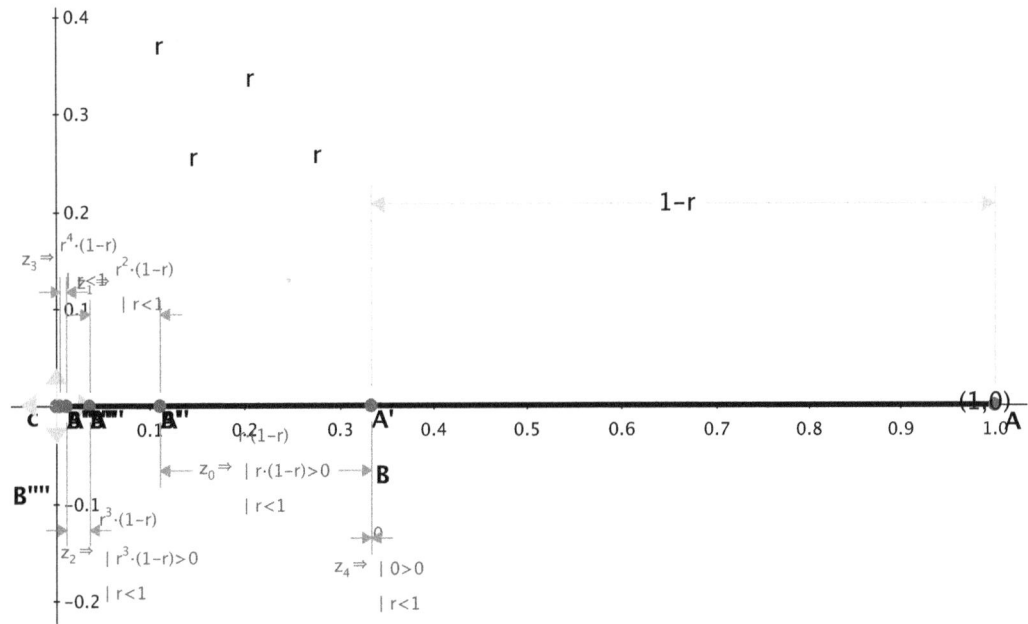

20.

Jump #	Distance jumped	Distance from destination	Total distance traveled
start	0	1	0
1	$1-r$	r	$1-r$
2	$r(1-r)$	r^2	$(1-r)+r(1-r)$
3	$r^2(1-r)$	r^3	$(1-r)+r(1-r)+r^2(1-r)$
4	$r^3(1-r)$	r^4	$(1-r)+r(1-r)+r^2(1-r)+r^3(1-r)$

21. $(1-r)+r(1-r)+r^2(1-r)+r^3(1-r)+r^4(1-r)+...$

If your students are familiar with sigma notation, or you use this as an opportunity to introduce them to it, the series would be:

$$\sum_{n=0}^{\infty} r^n(1-r)$$

22.
$(1-r)+r(1-r)+r^2(1-r)+r^3(1-r)+r^4(1-r)+...=1$

23. $(1-r)\left[1+r+r^2+r^3+r^4+...\right]=1$

24.
$$\frac{(1-r)\left[1+r+r^2+r^3+r^4+...\right]}{1-r}=\frac{1}{1-r}$$

$$1+r+r^2+r^3+r^4+...=\frac{1}{1-r}$$

Quick Questions

1. a. not geometric   b. geometric

   c. geometric   d. geometric

2. b. starting value = 1 r = -2, because the absolute value of r is greater than 1, this series does not converge to a finite sum, so the formula does not work. Think of it in terms of the frog. His first jump is MORE than the initial distance. We say it *diverges*.

   c. starting value = 4, common ratio is 1/3, sum = 6

   d. starting value = -2, common ratio is 2/5, sum = -10/3

3.

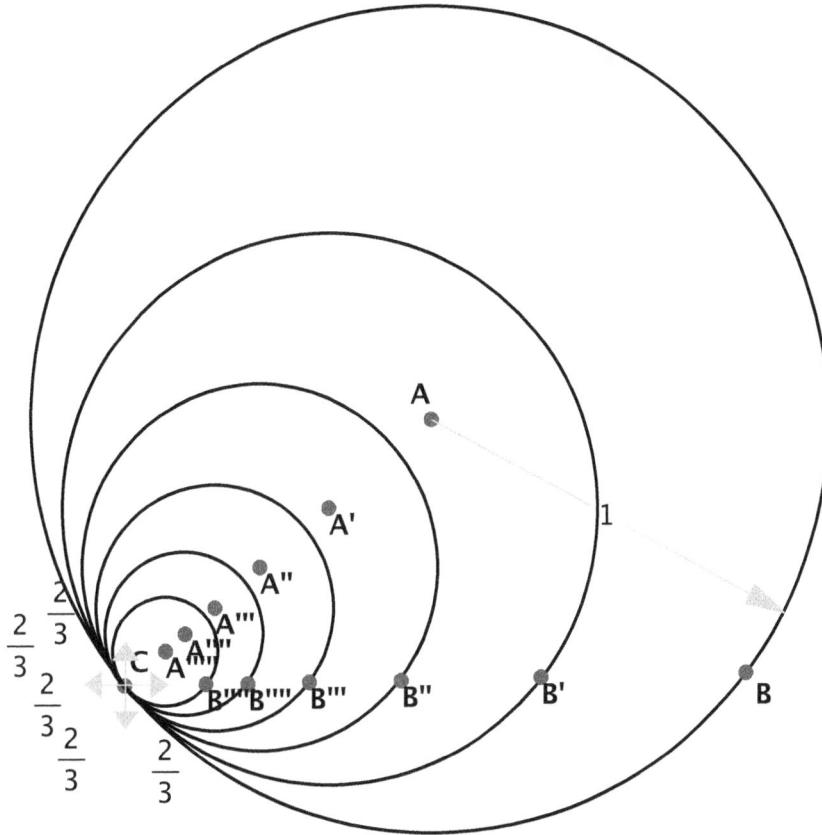

a. Write the infinite series of radii and calculate their sum.

$$1 + \frac{2}{3} + \frac{4}{9} + \frac{8}{27} + \dots$$

sum = 3

b. Write the infinite series of the areas and calculate their sum.

$$\pi + \frac{4}{9}\pi + \frac{16}{81}\pi + \frac{64}{729}\pi + \dots$$

sum = $\dfrac{9\pi}{5}$

Decimal Diversion

Fraction	Decimal Equivalent
$\dfrac{1}{9}$	$.\overline{1}$
$\dfrac{2}{9}$	$.\overline{2}$
$\dfrac{3}{9} = \dfrac{1}{3}$	$.\overline{3}$
$\dfrac{4}{9}$	$.\overline{4}$

Quick Questions

1. The part without repeating lines is not part of the repeating pattern, so it should be added in separately:

$$1 + \left(.6 + .06 + .006 + ...\right) = 1 + \frac{.6}{1 - .1}$$

$$= 1 + \frac{.6}{.9}$$

2.

$$.03 + .0003 + .000003 + .00000003 + ... = \frac{.03}{1 - .01}$$

$$= \frac{.03}{.99}$$

3.

$$.1 + (.09 + .009 + .0009 + ...) = .1 + \frac{.09}{1 - .1}$$

$$.1 + \frac{.09}{.9} = 1.1$$

4.

$$.123 + .000123 + .000000123 + ... = \frac{.123}{1 - .001}$$

$$\frac{.123}{.999}$$

Lesson Three

Objectives

The student will be able to use the coordinate graphing capabilities of Geometry Expressions to create a two dimensional representation of Zeno's paradox.

The student will be able to recognize and use connections among mathematical ideas.

Lesson Notes

In this lesson the student creates a graph to illustrate Zeno's paradox.

4. The lines appear to intersect, which should indicate that Achilles does indeed catch the tortoise.

7. **a units of distance**

8. **a units of time**. Achilles equation is $y = x$, so his distance and time are equivalent.

10. **ab units**. The tortoise's position is given by the equation **a + bx**. The time interval that has transpired is a units, so replacing a for x gives: **a + ba**. The First **a** is the initial position of the tortoise, so the new distance traveled is just **ba**.

11.

12. **Distance covered by Achilles: ab units**

Time for Achilles to cover that distance: ab units

Distance covered by Tortoise in that time: ab^2 units

This last answer is a good algebraic exercise. This time the x-coordinate is a + ab from the information above. So, by substituting and simplifying:

$$a + bx$$

$$a + b(a + ab)$$

$$a + ab + ab^2$$

The first two terms are the previous distance, so the final term is the new distance covered.

13.

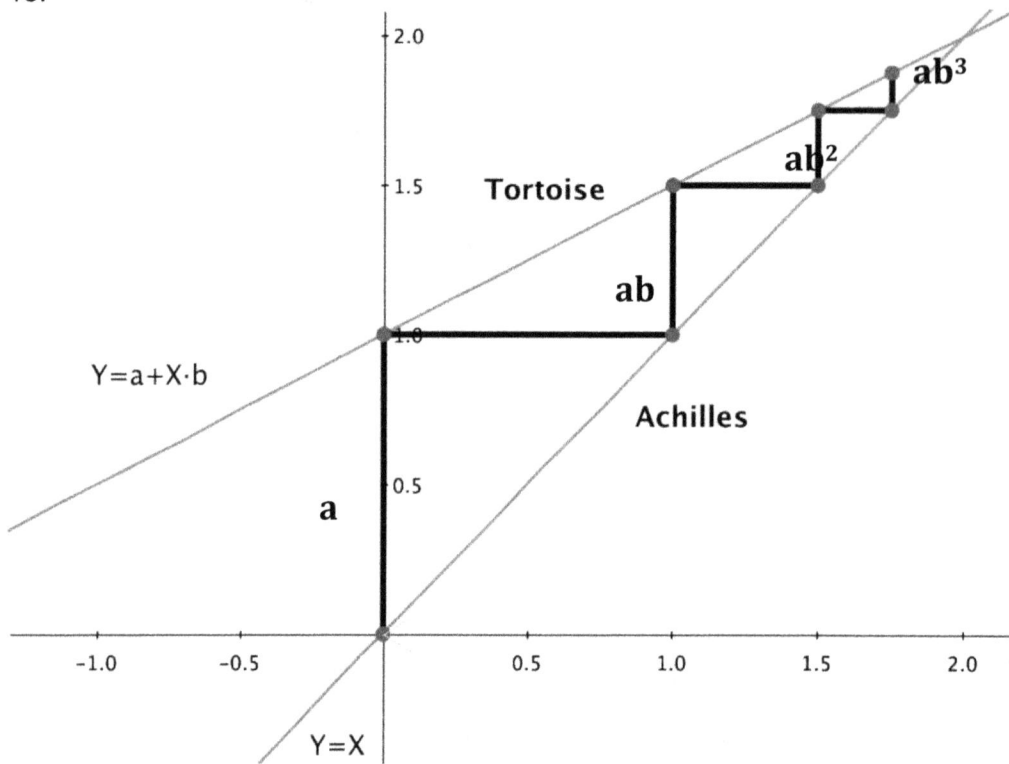

14. **Total Distance (4 time periods):** $a + ab + ab^2 + ab^3$

15. **Next term:** ab^4

16. **Starting term:** a **Common Ratio:** b

17. **Sum:** $\dfrac{a}{1-b}$

18.

$$z_5 \Rightarrow \quad a\cdot b^3$$
$$| \; a\cdot b^3 > 0$$

$$M \quad z_6 \Rightarrow \left(\dfrac{a}{1-b}, \dfrac{a}{1-b} \right)$$

2.0

L

J K

$$z_4 \Rightarrow \quad a\cdot b^2$$
$$| \; a > 0$$

H

1.5   Tortoise

$$a\cdot b$$
$$z_2 \Rightarrow \quad | \; a > 0$$
$$| \; b > 0$$

$$z_1 \Rightarrow \quad a$$
$$| \; a > 0$$

E   G
1.0

Y=a+X·b

Achilles

$$z_0 \Rightarrow \quad a$$
$$| \; a > 0$$ 0.5

$$z_3 \Rightarrow \quad a\cdot b$$
$$| \; a\cdot b > 0$$

−1.0   −0.5   F   0.5   1.0   1.5   2.0   2.5

Y=X

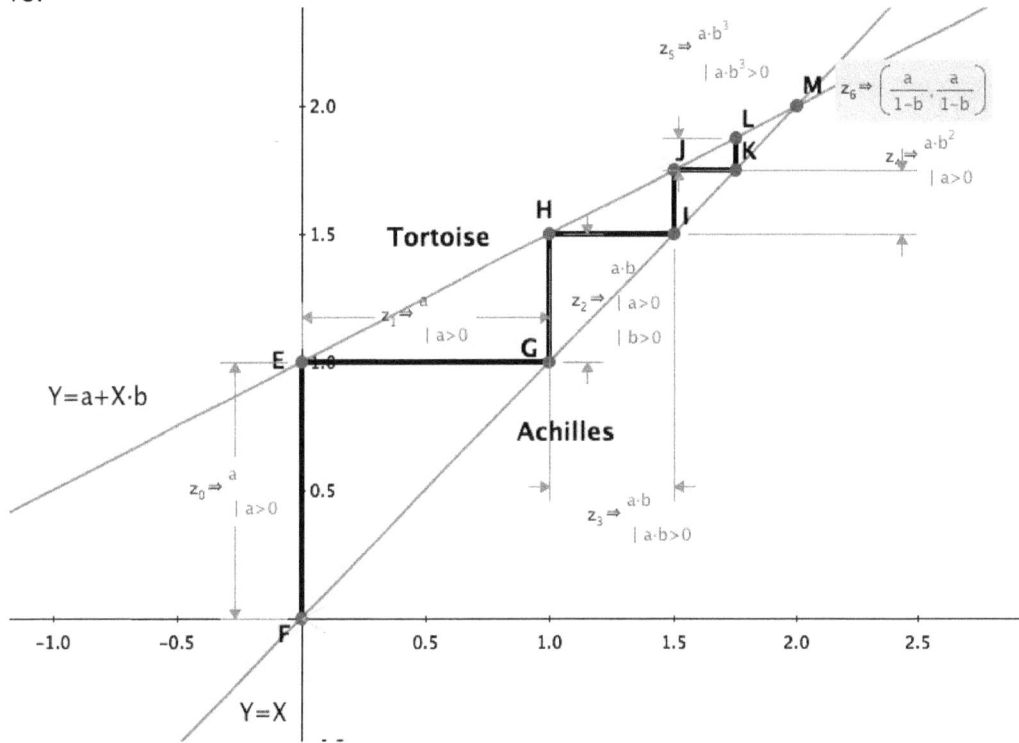

Quick Questions

2. The intersection point moves further and further away (in both time and distance)

3. You are increasing the speed of the Tortoise relative to Achilles speed. So, it takes longer for Achilles to close the gap between them.

4. The lines are parallel.

5. The Tortoise is moving at the same speed as Achilles and there is no paradox, Achilles will never catch up.

6. The Tortoise is actually moving faster than Achilles and the distance between them will grow without bound. In terms of a geometric series, the common ratio is greater than one, so the series diverges and has an infinitely large sum.

7. In terms of a geometric series, the common ratio is greater than one, so the series diverges and has an infinitely large sum. Because the Tortoise is travelling faster, he will increase the distance between them.

Interesting Investigation

Result when b is approximately -0.8

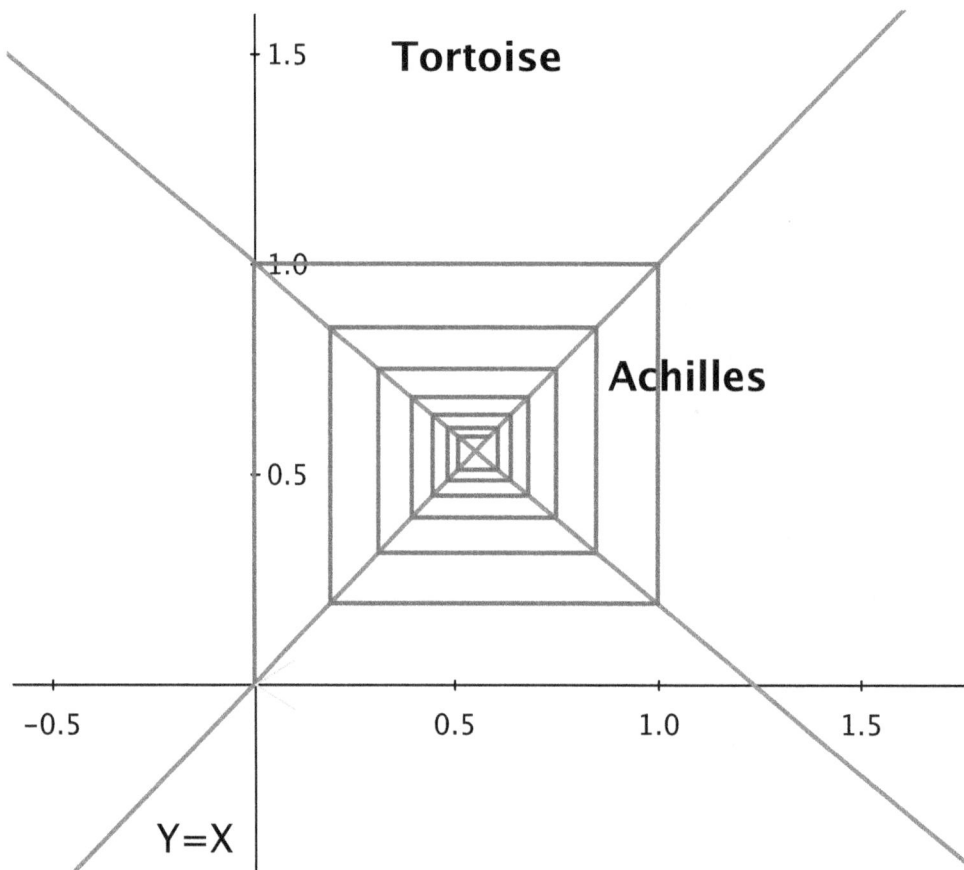

Result when b is approximately -1.1

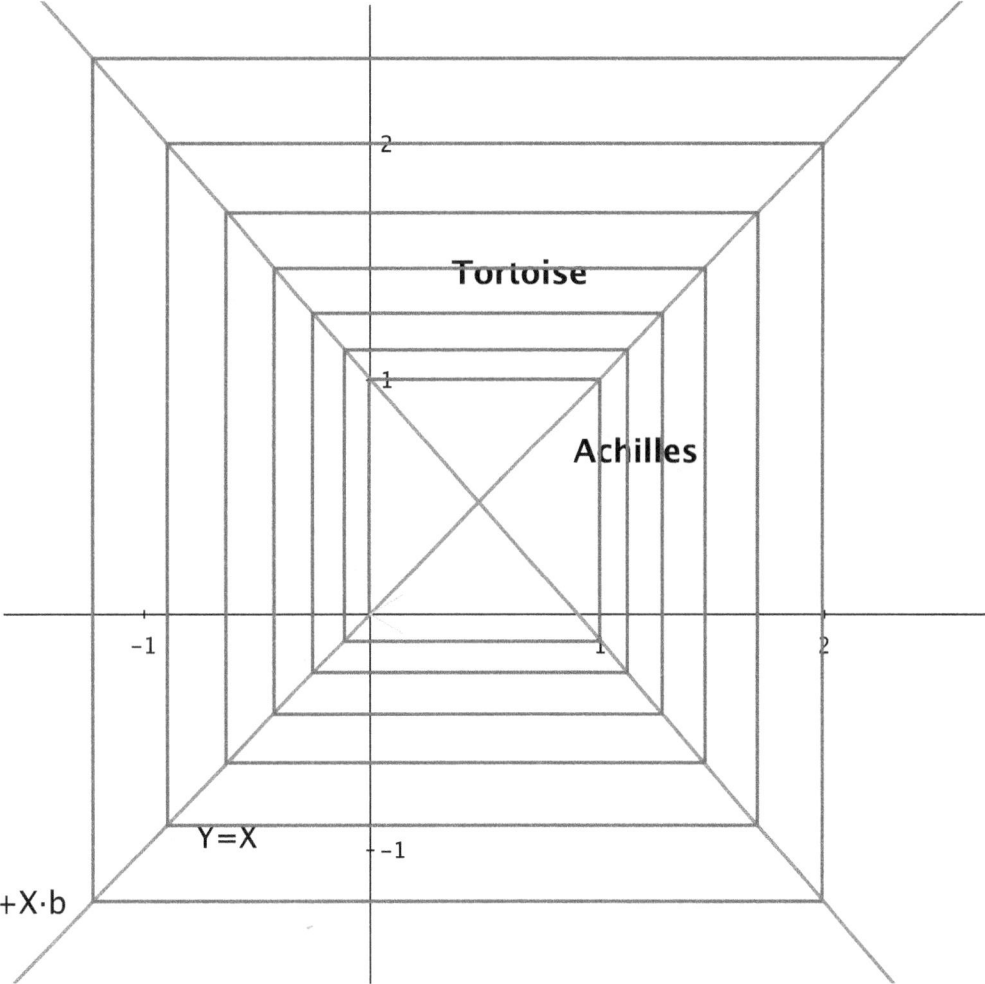

Tortoise

Achilles

Y=X

+X·b

2

1

−1

1

2

−1

Lesson Four

Objectives

The student will be able to use the basic commands of Geometry Expressions.

The student will be able to calculate areas between functions and segments using Geometry Expressions.

The student will be able to recognize and use connections among mathematical ideas.

The student will be able to use reasoning and methods of proof.

Lesson Notes

In this lesson the student uses a similar technique to Archimedes method with an absolute value function. The exercise gives the student some practice with areas and basic geometry using Geometry Expressions.

Quick Questions

1. In the way we've constructed our drawing, the value of x is negative, so –x is a positive number.

2 .

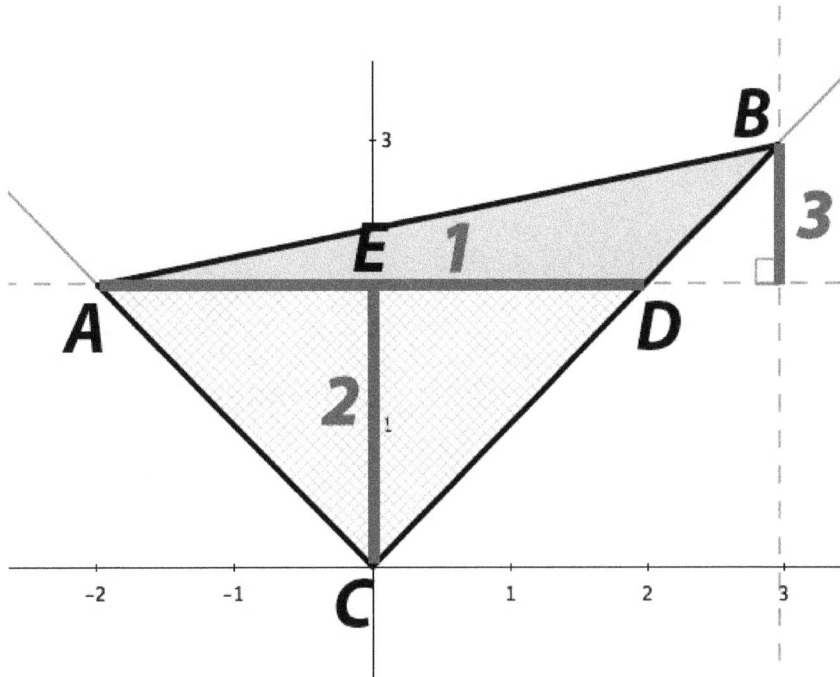

1 Segment **AD**. The distance from **A** to **E** is **–x**. The distance from **E** to **D** is also **–x**, so the total distance is **-2x**.

2 Segment **CE**. This is equivalent to the height of either point **A** or **D**, which must also be **–x**.

3 The height of triangle **ADB**. B has an x-coordinate of **h+x**, but the drawing is assuming that **h+x** is positive. The absolute value of a positive number is positive, or in this case, **h+x**. The length of the segment is the difference in height from the top to the bottom:

$$h + x - (-x)$$
$$h + 2x$$

3. Triangle ACD

$$\tfrac{1}{2}bh$$

$$\tfrac{1}{2}(-2x)(-x)$$

$$x^2$$

Triangle ABD

$$\tfrac{1}{2}bh$$

$$\tfrac{1}{2}(-2x)(h+2x)$$

$$-x(h+2x)$$

$$-hx-2x^2$$

AREA ACD:

$$x^2$$

AREA ABD:

$$-hx-2x^2$$

4. AREA ABC:

$$x^2+\left(-hx-2x^2\right)$$

$$-hx-x^2$$

6.

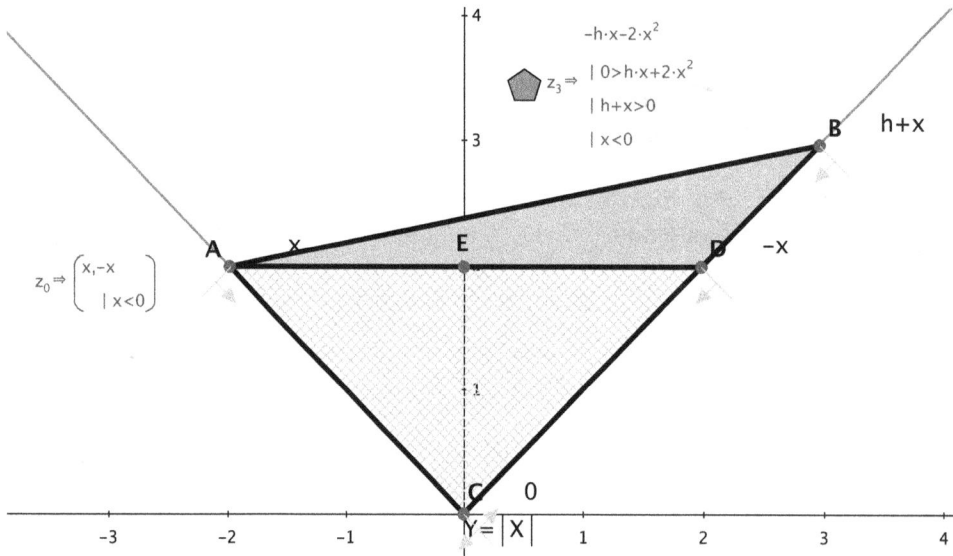

Lesson Five

Objectives

The student will be able to use Geometry Expressions to model Archimedes' process of quadrature.

The student will be able to use an infinite geometric series to approximate the area in a parabolic segment.

The student will be able to recognize and use connections among mathematical ideas.

Lesson Notes

In this lesson students use Geometry Expressions to find the area inside a parabola. They are performing Achimedes' process of quadrature by adding more and more triangular segments.

Informative Instructions

4. $\dfrac{h}{2}$

Quick Questions

1. The area of the triangle is *less* than the area between the segment and the parabola.

2.

Numeric area of ABC	Symbolic area of ABC
6.057598	$\dfrac{h^3}{8}$

5. A (x, x^2) B $\left(h+x, (h+x)^2\right)$

7. $(h+x)^2$

$h^2 + 2hx + x^2$

8. **x-values:**

$$\dfrac{x+\left(h+x\right)}{2}$$

$$=\dfrac{h+2x}{2}$$

$$=\dfrac{h}{2}+x$$

y-values:

$$\dfrac{x^2+\left(h+x\right)^2}{2}$$

$$=\dfrac{x^2+h^2+2hx+x^2}{2}$$

$$=\dfrac{h^2+2hx+2x^2}{2}$$

$$\dfrac{h^2}{2}+hx+x^2$$

9.

	Triangle ADC	Triangle BDC
Height	$\dfrac{h}{2}$	$\dfrac{h}{2}$
Base	$\dfrac{h^2}{4}$	$\dfrac{h^2}{4}$
Area	$\dfrac{1}{2}\cdot\dfrac{h}{2}\cdot\dfrac{h^2}{4}=\dfrac{h^3}{16}$	$\dfrac{h^3}{16}$

TOTAL AREA: $2 \cdot \dfrac{h^3}{16} = \dfrac{h^3}{8}$

The most involved calculation here is the base, which is length **CD**. We calculated the height of point **D** in the previous question, so subtracting the height of point **C** will give the length of this segment.

$$\frac{h^2}{2} + hx + x^2 - \left(\frac{h}{2} + x\right)^2$$

$$\frac{h^2}{2} + hx + x^2 - \left(\frac{h^2}{4} + 2 \cdot \frac{hx}{2} + x^2\right)$$

$$\frac{h^2}{2} + hx + x^2 - \frac{h^2}{4} - hx - x^2$$

$$\frac{h^2}{2} - \frac{h^2}{4}$$

$$\frac{h^2}{4}$$

10. It is amazingly important that students realize, either on their own, or with gentle prodding from you, two important aspects of what they have just done:

    1. The area of each of the smaller triangles is the same.

    2. The area of the triangle is independent of the value of (the location of the point along the parabola).

Archimedes, again

2. The coordinates should be: $\dfrac{h}{4} + x$ and $\dfrac{3h}{4} + x$ respectively because they occur at the quarter positions.

6. AREA of triangle **ACF**: $\dfrac{h^3}{64}$ AREA of triangle **ACE**: $\dfrac{h^3}{64}$

7. **TOTAL NEW AREA:** $\dfrac{h^3}{32}$

8. **RATIO:** $\dfrac{\dfrac{h^3}{32}}{\dfrac{h^3}{8}} = \dfrac{8}{32} = \dfrac{1}{4}$

11.   **AREA OF EACH NEW TRIANGLE:** $\dfrac{h^3}{512}$

TOTAL AREA of 4 NEW TRIANGLES: $\dfrac{h^3}{128}$

$\dfrac{\text{4 NEW}}{\text{2 PREVIOUS}}$: $\dfrac{\dfrac{h^3}{128}}{\dfrac{h^3}{512}} = \dfrac{512}{128} = \dfrac{1}{4}$

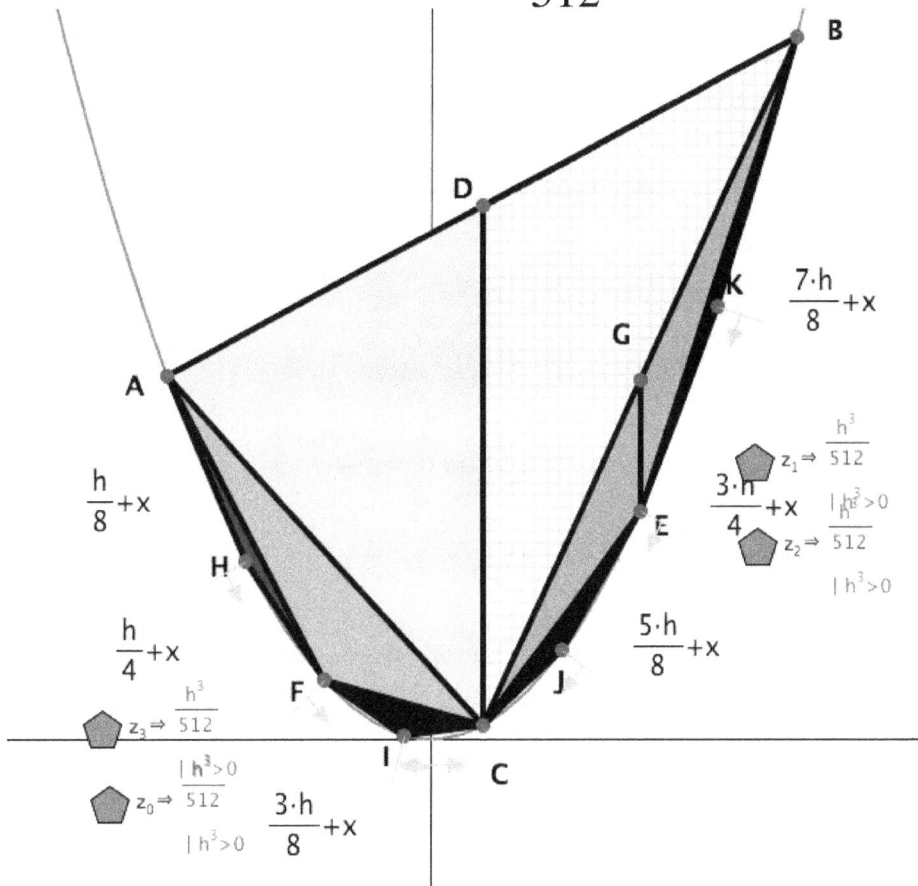

$\dfrac{h}{8}+x$

$\dfrac{h}{4}+x$

$z_3 \Rightarrow \dfrac{h^3}{512}$

$\mid h^3 > 0$

$z_0 \Rightarrow \dfrac{h^3}{512}$

$\mid h^3 > 0$

$\dfrac{3 \cdot h}{8}+x$

$\dfrac{5 \cdot h}{8}+x$

$\dfrac{3 \cdot h}{4}+x$

$z_1 \Rightarrow \dfrac{h^3}{512}$

$\mid h^3 > 0$

$z_2 \Rightarrow \dfrac{h^3}{512}$

$\mid h^3 > 0$

$\dfrac{7 \cdot h}{8}+x$

12. $\dfrac{h^3}{8} + \dfrac{h^3}{32} + \dfrac{h^3}{128} + \dfrac{h^3}{512} + \ldots + \dfrac{h^3}{2^{2n+1}}$

or, for students familiar with sigma notation:

$$\sum_{n=1}^{\infty} \dfrac{h^3}{2^{2n+1}}$$

13. Initial Term: $\dfrac{h^3}{8}$ Common Ratio: $\dfrac{1}{4}$

Sum: $\dfrac{\dfrac{h^3}{8}}{1 - \dfrac{1}{4}} = \dfrac{h^3}{6}$

Actual Archimedes

Students may complete this exercise as an optional activity.
Note a couple of differences in order to construct this drawing:

1) In place of a function, you'll have to draw a parabola ⌣ , select it
   and **Constrain** the **Implicit equation** to Y=X².
2) The **Point proportional** constraint works differently for functions than
   for the parabola, so in order to get the same area result, use vertical
   lines to constrain the position of points C and D on the parabola.
   Position the lines making them perpendicular to the x axis and

   Constrain the **Coordinates** of their intersection points with the
   x axis, then constrain points C and D on the parabola to be incident
   to these lines.

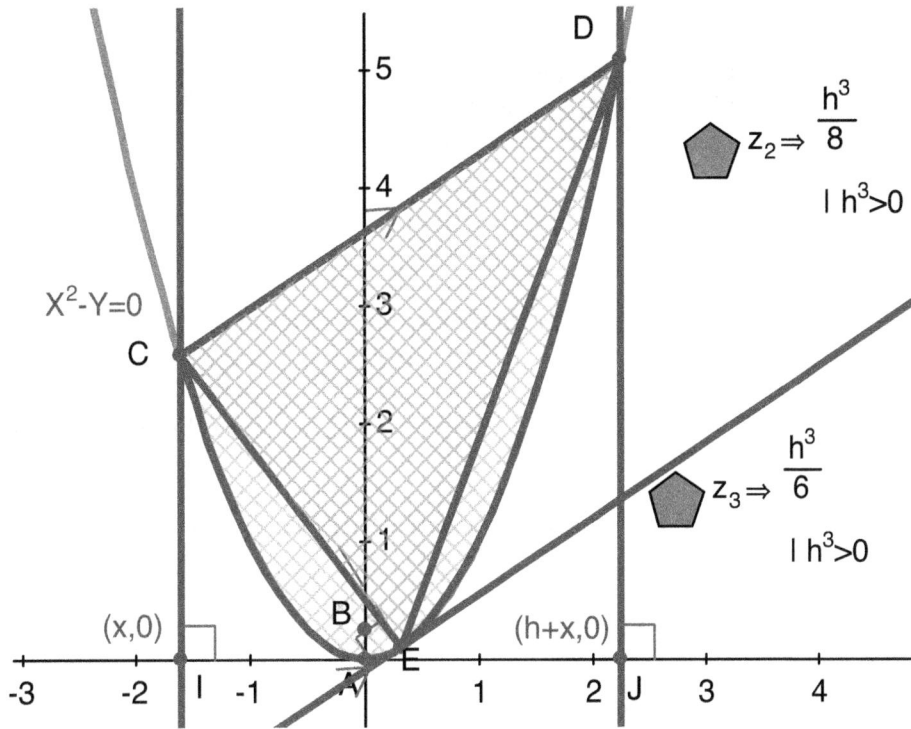

$X^2-Y=0$

C

D

$(x,0)$

$(h+x,0)$

B

E

I

A

J

-3 -2 -1 1 2 3 4

5 4 3 2 1

$z_2 \Rightarrow \dfrac{h^3}{8}$

$|\ h^3>0$

$z_3 \Rightarrow \dfrac{h^3}{6}$

$|\ h^3>0$

113

Lesson Six

Objectives

The student will be able to use Geometry Expressions to model Archimedes' process of quadrature.

The student will be able to use an infinite geometric series to approximate the area in a parabolic segment.

The student will be able to recognize and use connections among mathematical ideas.

Lesson Notes

In this lesson, students again use Geometry Expressions to model the quadrature process. However, this time they will start with a triangle to over-approximate the area of the parabolic segment and use the iterative process to subtract the extra areas.

Informative Instructions

6. $\dfrac{h^3}{4}$ 8. $\dfrac{h^3}{16}$ 9. $\dfrac{\dfrac{h^3}{16}}{\dfrac{h^3}{4}} = \dfrac{1}{4}$ 10. $\dfrac{h^3}{4} - \dfrac{h^3}{16} = \dfrac{3h^3}{16}$

11. When the drawings get "busy" it is sometimes difficult to select the appropriate object. Remind students that they can zoom in and out on different parts of the diagram. They can also select the curve *anywhere* along the curve, so select it in a spot where there is not an abundance of other objects. Also, it is always a good idea to remind them about **CNTRL-Z (UNDO)**, which will help them out of *a lot* of mistakes.

12. a. $\dfrac{h^3}{128}$ b. $\dfrac{2h^3}{128} = \dfrac{h^3}{64}$ c. $\dfrac{1}{4}$

d. $\dfrac{h^3}{4} - \dfrac{h^3}{16} - \dfrac{h^3}{64} = \dfrac{11h^3}{64}$

e. 4 f. $\dfrac{h^3}{1024}$ g. $\dfrac{4h^3}{1024} = \dfrac{h^3}{256}$

h. When you zoom in close, it may not be possible to select the parabola. To work around this, select the area of the original large triangle **ABC**, and choose **CNTRL-H** to hide it.

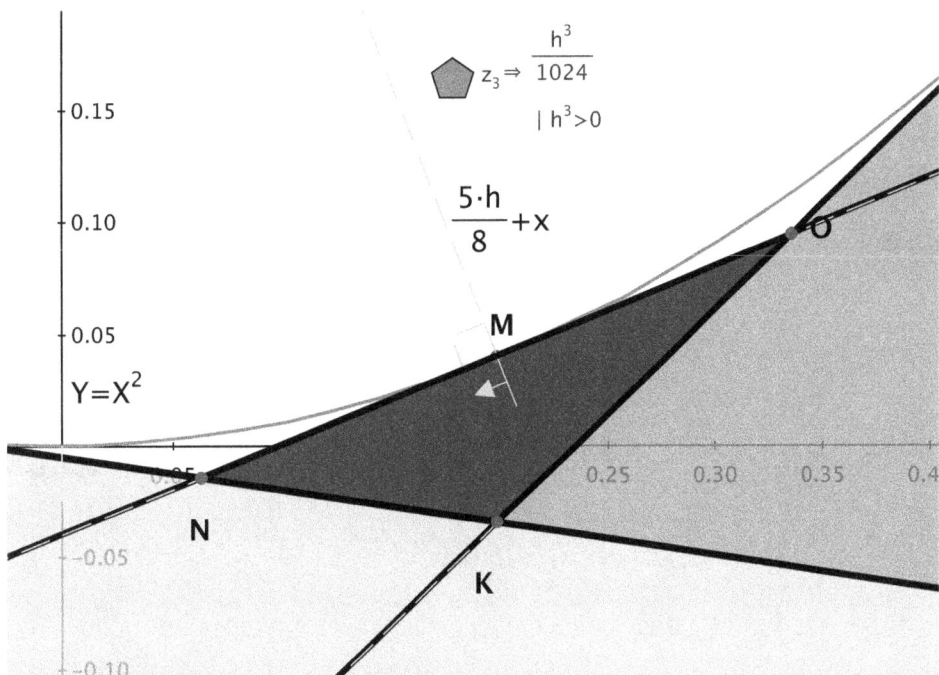

i. $\dfrac{h^3}{4} - \dfrac{h^3}{16} - \dfrac{h^3}{64} - \dfrac{h^3}{256} - \cdots \dfrac{h^3}{4^n}$

j. This series is not a sum, but a difference, so in order to apply the formula, we need to adjust it a bit:

$$\frac{h^3}{4} - \frac{h^3}{16} - \frac{h^3}{64} - \frac{h^3}{256} - \cdots$$

$$\frac{h^3}{4} + \left(-\frac{h^3}{16} - \frac{h^3}{64} - \frac{h^3}{256} - \cdots \right)$$

$$\frac{h^3}{4} - \left(\frac{h^3}{16} + \frac{h^3}{64} + \frac{h^3}{256} + \cdots \right)$$

Now think of the expression in the parenthesis as a geometric series and apply the formula using a starting value of $\dfrac{h^3}{16}$ with a common ratio of $\dfrac{1}{4}$.

Subtract the resulting sum from the original term to find the answer.

$$\frac{\dfrac{h^3}{16}}{1 - \dfrac{1}{4}} = \frac{\dfrac{h^3}{16}}{\dfrac{3}{4}} = \frac{h^3}{12}$$

$$\frac{h^3}{4} - \frac{h^3}{12} = \frac{2h^3}{12} = \frac{h^3}{6}$$

Lesson Seven

Objectives

The students will be able to use Geometry Expressions to find the area under a given function.

The student will be able to recognize and use connections among mathematical ideas.

The student will be able to use reasoning and methods of proof.

Lesson Notes

In this lesson, students are introduced to one of the fundamental tasks of modern calculus, finding the area under a curve.

Informative Instructions

7. $\dfrac{1}{2} h \left(b_1 + b_2 \right)$

8.

Object	Length/Area
BC (b_1 of trapezoid)	$(h+t)^2 = h^2 + 2ht + t^2$
AD (b_2 of trapezoid)	t^2
DC (height of trapezoid)	h
Area of trapezoid	$\frac{1}{2}h\left(h^2 + 2ht + t^2 + t^2\right)$ $\frac{1}{2}h\left(h^2 + 2ht + 2t^2\right)$ $\frac{h^3}{2} + h^2t + ht^2$
Area between AB and parabola	$\frac{h^3}{6}$
Area below parabola	$\dfrac{h^3}{2} + h^2t + ht^2 - \dfrac{h^3}{6}$ $\dfrac{h^3}{3} + h^2t + ht^2$

This is the formula for a definite integral that students learn in calculus.

1. $$\frac{(5)^3}{3} + (5)^2(-2) + (5)(-2)^2$$

$$= \frac{35}{3} \approx 11.6667$$

2. $$\frac{(5)^3}{3} + (5)^2(0) + (5)(0)^2$$

$$= \frac{125}{3} \approx 41.6667$$

3. The area would continue to grow larger, without bound.

4. Assume x represents the width of the interval. In this case, because we are starting at 0, the width and the upper bound are the same.

$$\frac{x^3}{3} + x^2(0) + x(0) = \frac{35}{3}$$

$$\frac{x^3}{3} = \frac{35}{3}$$

$$x^3 = 35$$

$$x = \sqrt[3]{35}$$

$$x \approx 3.27107$$

Exciting Extension

***Usually, extensions are more advanced topics to differentiate learning for your top students. This is not the case here. Don't skip this extension!!! This is where we show the students that GX will now calculate the area for us automatically. We will use this to revisit all the work of the previous lessons.

5. $\dfrac{h^3}{3} + h^2t + ht^2$

6.

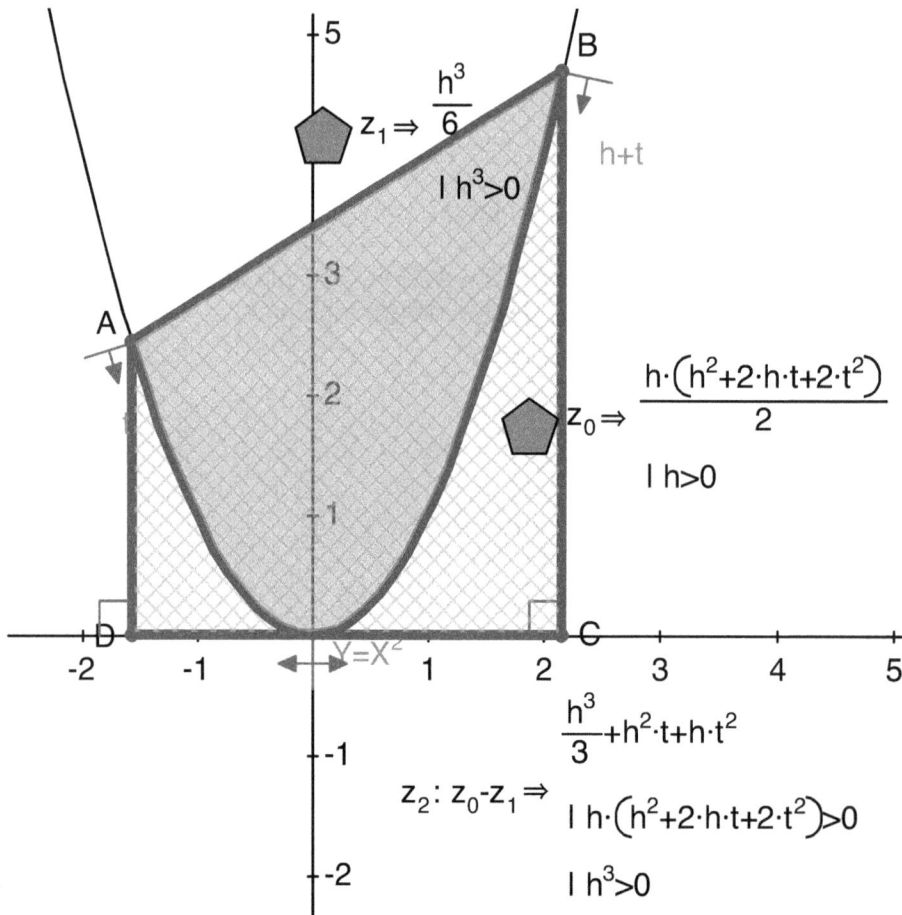

7. h

8. $\dfrac{h^3}{3}$

9.

Lesson Eight

Objectives

The student will use Geometry Expressions to approximate the area under a square root function.

The student will be able to recognize and use connections among mathematical ideas.

The student will be able to use reasoning and methods of proof.

Lesson Notes

This lesson brings together the previous lessons to investigate the area under a square root function.

Interesting Investigation

6. The area should be exactly half of the original area.

7.   a. Point B (t, t^2)

       b. Point C (t^2, t)

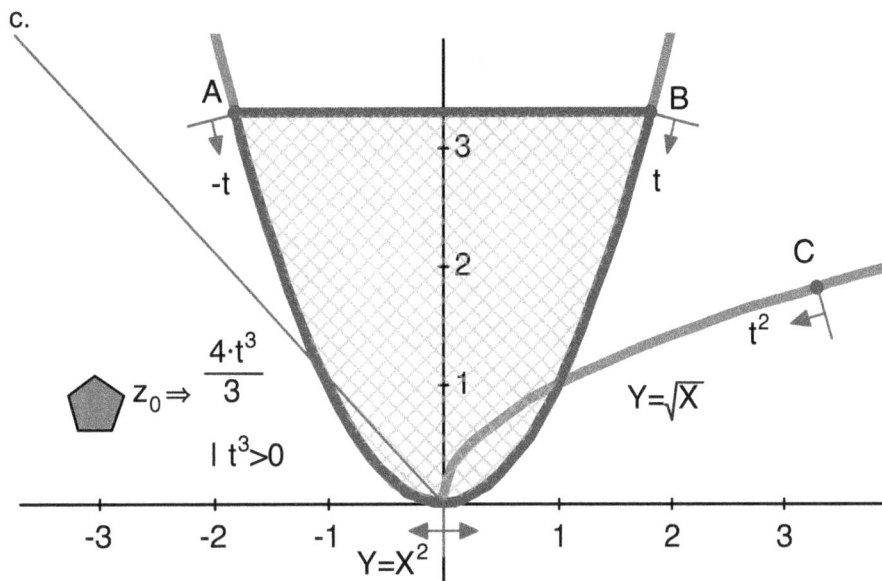

c.

A

B

$-t$

t

$+3$

$+2$

C

$z_0 \Rightarrow \dfrac{4 \cdot t^3}{3}$

$+1$

t^2

$Y=\sqrt{X}$

$\mid t^3 > 0$

-3 -2 -1 1 2 3

$Y=X^2$

d. $\dfrac{4t^3}{3} \div 2 = \dfrac{2t^3}{3}$

7. When constructing the arc, students may have trouble drawing the arc on the square root function, as opposed to the original function. To avoid this, click on the origin, then when you drag the mouse to point **C**, drag it **below** the x-axis until it is clear of the parabola, then swing up to point **C**.

8. $\dfrac{2u^{\frac{3}{2}}}{3}$

If the x-coordinate is u, then the y value is $\sqrt{u}$, substitute this for t in the area formula:

$$\frac{4t^3}{3} = \frac{4(\sqrt{u})^3}{3} = \frac{4(u^{\frac{1}{2}})^3}{3} = \frac{4u^{\frac{3}{2}}}{3}$$

And since we are measuring half of the area:

$$\frac{4u^{\frac{3}{2}}}{3} \div 2 = \frac{2u^{\frac{3}{2}}}{3}$$

Lesson Nine

Objectives

The student will use Geometry Expressions to generalize the area under a curve for other power functions.

The student will be able to recognize and use connections between mathematical ideas.

The student will be able to use reasoning and methods of proof.

The student will be able to make and investigate mathematical conjectures.

Lesson Notes

In this lesson, we take the work from the previous lesson on square root functions and generalize it to other power functions.

Quick Question

$$\tfrac{1}{2}bh = \tfrac{1}{2}t \cdot t = \frac{t^2}{2}$$

Informative Instructions

3. $\frac{t^3}{2}$

4. $\frac{t^4}{2}$

5. It appears as if the area of the triangle is $\frac{t^{b+1}}{2}$ for any power function $y=ax^b$

For example, for $y=ax^6$, the area of the triangle should be $\frac{t^7}{2}$.

7.

Function	Area of triangle	Area under curve
$y=x$	$\frac{t^2}{2}$	$\frac{t^2}{2}$
$y=x^2$	$\frac{t^3}{2}$	$\frac{t^3}{3}$
$y=x^3$	$\frac{t^4}{2}$	$\frac{t^4}{4}$
$y=x^4$	$\frac{t^5}{2}$	$\frac{t^5}{5}$
$y=x^5$	$\frac{t^6}{2}$	$\frac{t^6}{6}$
$y=x^6$	$\frac{t^7}{2}$	$\frac{t^7}{7}$
$y=x^n$	$\frac{t^{n+1}}{2}$	$\frac{t^{n+1}}{n+1}$

Quick Questions

1. a. $f(x) = x^{\frac{1}{2}}$ b. $\dfrac{x^{\frac{3}{2}}}{\frac{3}{2}} = \dfrac{2x^{\frac{3}{2}}}{3}$

2. The area will increase as the power increases.

3.

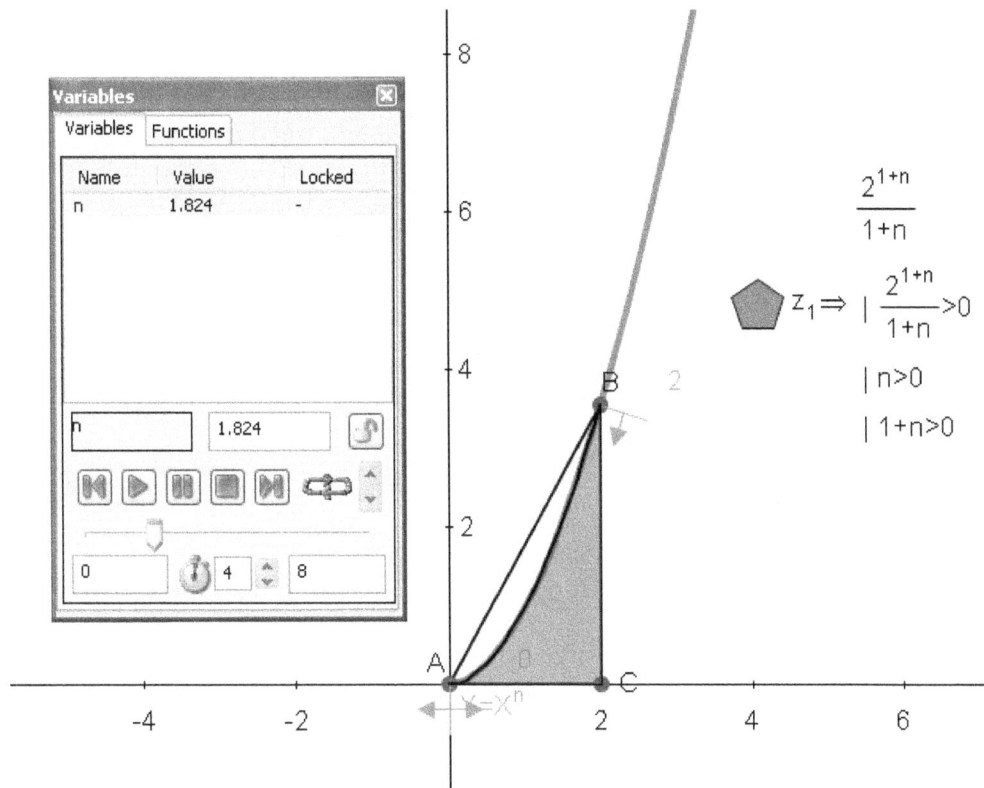

$\dfrac{2^{1+n}}{1+n}$

$z_1 \Rightarrow \ \vert \ \dfrac{2^{1+n}}{1+n} > 0$

$\vert \ n > 0$

$\vert \ 1+n > 0$

4. Because 1 stays fixed on the function, the area will actually decrease as the curve gets steeper.

5. The area decreases more dramatically in this example as the curve flattens out more quickly for a larger portion of the interval from 0 to 1.

Lesson Ten

Objectives

The student will use Geometry Expressions to graph and calculate the area under the normal curve between two particular values.

The student will be able to recognize and use connections between mathematical ideas.

Lesson Notes

In this lesson, students apply the ideas developed in previous lessons to the normal distribution in statistics.

Informative Instructions

Quick Questions:

1. It changes the center of the curve.

2. It changes how widely spread the curve is as well as the height of the center.

13.

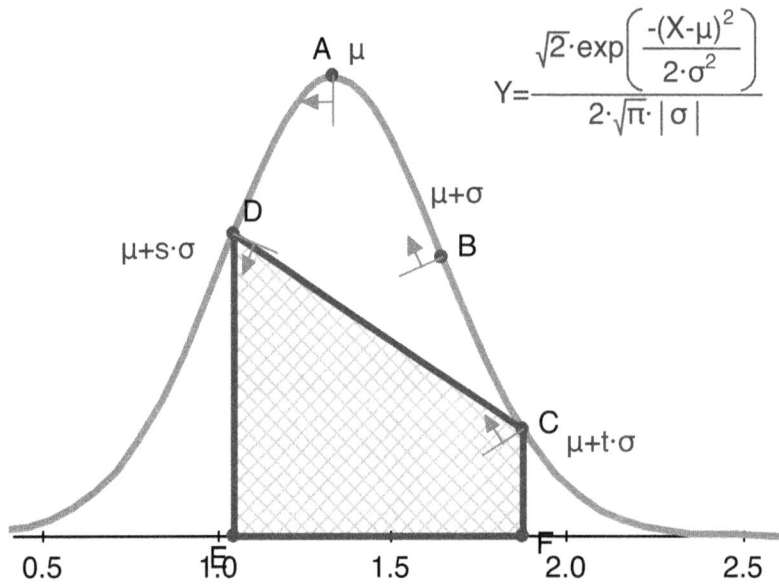

$$Y=\frac{\sqrt{2}\cdot\exp\left(\dfrac{-(X-\mu)^2}{2\cdot\sigma^2}\right)}{2\cdot\sqrt{\pi}\cdot|\sigma|}$$

Assessing Areas

11. about 0.683

12. about 0.954

13. about 0.997

14. 68, 95, 99.7

www.ingramcontent.com/pod-product-compliance
Lightning Source LLC
Chambersburg PA
CBHW081508200326
41518CB00015B/2419